SpringerBriefs in Optimization

SpringerBriefs in Optimization showcases algorithmic and theoretical techniques, case studies, and applications within the broad-based field of optimization. Manuscripts related to the ever-growing applications of optimization in applied mathematics, engineering, medicine, economics, and other applied sciences are encouraged.

More information about this series at http://www.springer.com/series/8918

Anatoly Zhigljavsky • Antanas Žilinskas

Bayesian and High-Dimensional Global Optimization

 Springer

Anatoly Zhigljavsky
Mathematics Institute
Cardiff University
Cardiff, UK

Antanas Žilinskas
Institute of Data Science
and Digital Technologies
Vilnius University
Vilnius, Lithuania

ISSN 2190-8354   ISSN 2191-575X   (electronic)
SpringerBriefs in Optimization
ISBN 978-3-030-64711-7   ISBN 978-3-030-64712-4   (eBook)
https://doi.org/10.1007/978-3-030-64712-4

Mathematics Subject Classification: 90C26, 90C15, 90C59, 65K05, 62F15

This Springer imprint is published by the registered company Springer Nature Switzerland AG
The registered company address is: Gewerbestrasse 11, 6330 Cham, Switzerland

Preface

Global optimization (GO) is a broad and active field of research including mathematical analysis of problems, development of algorithms and software, and applications to various real-word problems. Many approaches have been proposed to tackle different subclasses of GO problems. In this book we consider the algorithms aimed at two challenging classes of GO problems usually specified as black-box expensive and high-dimensional. We mainly focus on two approaches: the Bayesian approach and high-dimensional random or pseudo-random search.

Interests in Bayesian and high-dimensional GO have been growing fast during last few decades. The development of algorithms was stimulated by new applications especially in optimal engineering design where optimization problems are characterized as black-box and expensive. Important, recently emerged application areas are related to computer experiments and machine learning. The theoretical investigation in Bayesian and high-dimensional GO was carried out in parallel with the development of algorithms. In the present monograph, some of the original ideas are presented as well as their further developments, discussing also challenges and unsolved problems.

The book has three chapters. Chapter 1 starts by considering properties of high-dimensional cubes and balls, then various aspects of uniformity and space-filling are discussed. It is demonstrated that good uniformity of a set of points does not have to imply its good space-filling. Special attention in Chap. 1 is devoted to covering, which is a very important criterion of space-filling. In particular, the concept of weak covering is introduced, where only a large part of a feasible domain has to be covered by the balls with centers at given points, rather than the full domain as in the standard covering. In Chap. 2, we present our recent work in Bayesian approach to continuous non-convex optimization. A brief review precedes the main results to have our work presented in the context of challenges of the approach. Chapter 3 is devoted to global random search (GRS). It is not the aim of this chapter to cover the whole subject; we only make some potentially important notes on algorithms of GRS in continuous problems, mostly keeping in mind the use of such algorithms in reasonably large dimensions.

In the process of writing the book, we have spent long hours discussing its various parts, but responsibility for different chapters is not shared. Anatoly takes full responsibility for Chaps. 1 and 3 while Antanas is responsible for the material of Chap. 2.

Overall, the book is written for a wide circle of readers and will hopefully be appreciated by those interested in theoretical aspects of global optimization as well as practitioners interested mostly in the methodology. All those interested in applications of global optimization can also benefit from the book.

We are very grateful to both referees of the book for very detailed and very helpful comments. We are also very grateful the editors of the *SpringerBriefs in Optimization* and especially to Prof. Panos Pardalos for his encouragement of our project. We thank Jonathan Gillard, Jack Noonan, and Andrey Pepelyshev (Cardiff) for helping to improve English and maths in many parts of the book. The research by A. Žilinskas was funded by a grant (No. S-MIP-17-61) from the Research Council of Lithuania.

Cardiff, UK   Anatoly Zhigljavsky
Vilnius, Lithuania   Antanas Žilinskas
October 2020

Contents

1   **Space-Filling in High-Dimensional Sets** 1
  1.1   Peculiarities of High-Dimensional Sets 1
      1.1.1   Volume of the Unit Ball and the Ball of Unit Volume 1
      1.1.2   Concentration of Mass in the Unit Cube 3
      1.1.3   Almost All the Volume Is Near the Boundary................. 3
      1.1.4   Uniform i.i.d. Random Points Inside a Ball 5
      1.1.5   Distance Between Two Random Points in a Cube 6
      1.1.6   Mass Concentration for the Gaussian Distribution............ 7
      1.1.7   Volume of a Small Ball Placed In-Between 2^d
               Large Balls 7
      1.1.8   Volume of Intersection of a Cube and a Ball 9
  1.2   Space-Filling, Uniformity and Discrepancies 14
      1.2.1   Discrepancies 15
      1.2.2   Weighted Discrepancies 17
      1.2.3   Weighted Discrepancies for Farey Sequences and the RH.... 20
      1.2.4   Star-Discrepancy as Integration Error for Indicator
               Functions 21
      1.2.5   Other Characteristics of Space-Filling......................... 25
  1.3   Covering 29
      1.3.1   Covering Radius 29
      1.3.2   Star-Discrepancy and Covering Radius........................ 30
      1.3.3   Covering Radius and Weighted Discrepancy 31
      1.3.4   Weak Covering.. 32
      1.3.5   Relation Between Quantization and Weak Covering.......... 36
      1.3.6   Regularized Covering... 36
  1.4   Bibliographic Notes.. 37
  References 37

**2 Bi-objective Decisions and Partition-Based Methods in Bayesian
   Global Optimization** 41
  2.1 Original Algorithms... 42
  2.2 A Review of the Contemporary Methods 45
     2.2.1 Algorithms for Problems with Box Constraints 47
     2.2.2 Algorithms for Problems with Complicated Constraints 50
     2.2.3 Noisy Optimization 52
     2.2.4 Methods Using Derivatives..................................... 54
     2.2.5 Extensions to Large Dimensions 56
     2.2.6 Applications and What Publications Do Not Tell 58
  2.3 The Search Process as a Sequence of Rational Decisions............. 59
     2.3.1 Bi-objective Selection 64
     2.3.2 Numerical Example... 65
  2.4 Selection of a Probabilistic Model and Estimation
     of Its Parameters 69
     2.4.1 Gaussian Random Fields 69
     2.4.2 Approximating K_2 by $K_{2-\varepsilon}$ with Small ε 71
     2.4.3 Inverting the Covariance Matrix W_N in the Case $\gamma = 2$ 72
     2.4.4 MLE Estimation of μ and σ^2 for $\gamma = 2$.................... 72
     2.4.5 Numerical Experiments 73
  2.5 Partition-Based Algorithms.. 74
     2.5.1 Implementation 74
     2.5.2 Convergence 76
     2.5.3 Hybrid Algorithms.. 81
  References 82

3 Global Random Search in High Dimensions 89
  3.1 Main Ideas, Principles and Classes of Algorithms 89
     3.1.1 Main Principles and Important Classes of GRS
        Algorithms 90
     3.1.2 Multistart... 93
     3.1.3 Convergence and Rate of Convergence....................... 96
     3.1.4 Meta-heuristics in GRS.. 99
  3.2 Statistical Inference About f_* 101
     3.2.1 Statistical Inference in PRS: The Main Assumption 101
     3.2.2 Tail Index... 102
     3.2.3 Estimation of the Minimal Value of f 104
     3.2.4 Comparison of Random and Quasi-random Sequences....... 107
  3.3 Population-Based Algorithms 108
     3.3.1 Construction of Algorithms and Their Convergence.......... 108
     3.3.2 Homogeneous Transition Probabilities 111
  3.4 Bibliographic Notes.. 115
  References 116

Chapter 1
Space-Filling in High-Dimensional Sets

This chapter starts by considering, in Sect. 1.1, properties of high-dimensional cubes and balls; we will use many of these properties in other sections of this chapter and in Chap. 3. In Sect. 1.2, we discuss various aspects of uniformity and space-filling and demonstrate that good uniformity of a set of points is by no means implying its good space-filling. In Sect. 1.3, we consider space-filling from the viewpoint of covering and pay much attention to the concept of weak covering, where only a large part of a cube (or other set) has to be covered by the balls with centres at given points, rather than the full cube as in the standard covering. In Sect. 1.4, we provide bibliographic notes and give additional references.

1.1 Peculiarities of High-Dimensional Sets

In this section, we briefly illustrate several counter-intuitive facts related to high-dimensional cubes and balls. These facts quite often lead to creation of badly performing heuristics in multivariate optimization and misunderstanding of the behaviour of optimization algorithms in high-dimensional spaces.

1.1.1 Volume of the Unit Ball and the Ball of Unit Volume

Let

$$B_d(r) = \{x \in \mathbb{R}^d : \|x\| \leq r\} \tag{1.1.1}$$

be the d-dimensional ball with centre at 0 and radius r; here $\| \cdot \|$ is the usual Euclidean norm in $\mathbb{R}^d$. The volume of this ball is given by the formula

© The Author(s) 2021
A. Zhigljavsky, A. Žilinskas, *Bayesian and High-Dimensional Global Optimization*,
SpringerBriefs in Optimization, https://doi.org/10.1007/978-3-030-64712-4_1

$$\text{vol}(B_d(r)) = r^d V_d, \quad \text{where} \quad V_d = \text{vol}(B_d(1)) = \frac{\pi^{d/2}}{\Gamma(d/2+1)} \qquad (1.1.2)$$

and $\Gamma(\cdot)$ is the gamma function. The volumes V_d decrease very fast as d grows; see Fig. 1.1. For example, $V_{100} \simeq 2.368 \cdot 10^{-40}$. As $d \to \infty$,

$$V_d^{1/d} \simeq \frac{\sqrt{2\pi e}}{\sqrt{d}} + O\left(\frac{\log d}{d^{3/2}}\right). \qquad (1.1.3)$$

Consider now a ball of volume 1, that is, the ball $B_d(r_d)$ with r_d defined by $\text{vol}(B_d(r_d)) = 1$. From (1.1.3), the sequence of radii r_d grows proportionally to $\sqrt{d}$ as $d \to \infty$

$$r_d = \frac{\sqrt{d}}{\sqrt{2\pi e}} + O\left(\frac{1}{\sqrt{d}}\right), \qquad (1.1.4)$$

where $1/\sqrt{2\pi e} \simeq 0.242$. This is only about twice shorter than the half-diagonal (its length is $\sqrt{d}/2$) of the d-dimensional unit cube $[0, 1]^d$. Table 1.1 gives approximate values of r_d.

Summary *In high dimensions, the volume of the unit ball is extremely small and the radius of the ball of volume one is rather large.*

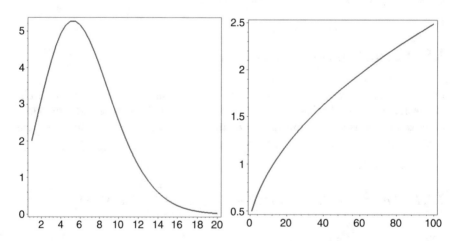

Fig. 1.1 Left: vol($B_d(1)$) as a function of d. Right: radius r_d of the ball with vol($B_d(r_d)$) = 1

Table 1.1 Radius of the ball of unit volume for various dimensions

d	1	2	3	4	5	6	7	8	9
r_d	0.5	0.564	0.62	0.671	0.717	0.761	0.8	0.839	0.876
d	10	20	30	40	50	100	200	500	1000
r_d	0.911	1.201	1.43	1.626	1.8	2.49	3.477	5.45	7.682

1.1.2   Concentration of Mass in the Unit Cube

Consider the cubes $C_\varepsilon = [\varepsilon/2, 1 - \varepsilon/2]^d$ with $0 \le \varepsilon < 1$. If $\varepsilon > 0$, we consider the cube C_ε as interior to the unit cube $C_0 = [0, 1]^d$, which is obtained by setting $\varepsilon = 0$. We clearly have $\mathrm{vol}(C_0) = 1$ and the volume of the interior cube $\mathrm{vol}(C_\varepsilon) = (1 - \varepsilon)^d$ tending to 0 (as $d \to \infty$) exponentially fast for any $\varepsilon \in (0, 1)$. If, as $d \to \infty, \varepsilon = \varepsilon_d$ tends to 0 slower than $1/d$, then the volume of C_ε still tends to 0. In particular, if $\varepsilon_d = c/d^{1-\delta}$ with $0 < \delta < 1$ and $c > 0$, then

$$\mathrm{vol}(C_\varepsilon) = \left(1 - \frac{c}{d^{1-\delta}}\right)^d \simeq \exp\{-cd^\delta\} \to 0, \quad d \to \infty.$$

Consider now a random vector $X = (x_1, \ldots, x_d)^\top$ uniformly distributed on the unit cube $[0, 1]^d$. Then $(x_1 - 1/2)^2, \ldots, (x_d - 1/2)^2$ are independent r.v. on $[a, b] = [0, 1/4]$ with mean $1/12$. The Hoeffding's inequality gives for any $\epsilon > 0$

$$\mathbb{P}\left\{\left| (x_1 - 1/2)^2 + \ldots + (x_d - 1/2)^2 - \frac{d}{12} \right| \ge \epsilon d\right\} \le 2e^{-2d\epsilon^2/(b-a)} = 2e^{-8d\epsilon^2}.$$

Therefore, most of the mass of the cube $[0, 1]^d$ is concentrated in the annulus around the sphere with radius $\sqrt{d/12}$ centred at the middle of the cube, the point $(1/2, \ldots, 1/2)^\top$.

Summary   *In high dimensions, the volume of the unit cube is concentrated near the boundary and at the same time very close to a sphere of radius $\sqrt{d/12}$. There is virtually no mass in the vicinity of cube's 2^d vertices. Despite the cube being convex, it can be thought of having the shape depicted in Fig.* 1.2.

1.1.3   Almost All the Volume Is Near the Boundary

A d-dimensional body is a measurable set $A \subset \mathbb{R}^d$ with $0 < \mathrm{vol}(A) < \infty$. For large d, the mass of a general d-dimensional body A, and not only of a cube, is concentrated near the boundary. Indeed, for any $0 \le \varepsilon < 1$, we can define the set $A_{1-\varepsilon} = \{(1 - \varepsilon)x : x \in A\}$. Then, by splitting A and $A_{1-\varepsilon}$ into infinitesimal cubes and adding up their volumes, we find

$$\mathrm{vol}(A_{1-\varepsilon}) = (1 - \varepsilon)^d \mathrm{vol}(A). \tag{1.1.5}$$

As an example, consider the balls $B_d(1)$ and $B_d(1 - \epsilon)$ defined by (1.1.1). The difference $B_d(1) \setminus B_d(1 - \epsilon)$ is the annulus; see Fig. 1.3. Using (1.1.2) we can compute the ratio of the volume of this annulus to the volume of the unit ball:

$$\frac{\mathrm{vol}\,[B_d(1) \setminus B_d(1 - \epsilon)]}{\mathrm{vol}(B_d(1))} = 1 - \varepsilon^d.$$

Fig. 1.2 Graphical
representation of a
d-dimensional cube

Fig. 1.3 Annulus of a unit
ball ($d = 2$)

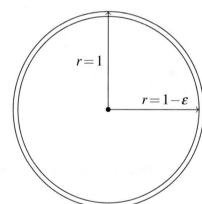

This ratio tends to 1 exponentially fast as $d \to \infty$. The ratio of volume of the ball
$B_d(1 - \epsilon)$ to the volume of the unit ball $B_d(1)$ is $(1 - \varepsilon)^d$, which is in full agreement
with (1.1.5).

Summary *Almost all the mass of a general high-dimensional body is concen-trated near the boundary.*

1.1.4 Uniform i.i.d. Random Points Inside a Ball

The norm of a uniformly distributed random point in $B_d(1)$ is a r.v. with density $p_d(r) = dr^{d-1}, 0 \le r \le 1$, and hence the probability that this norm exceeds t is $1 - t^d, 0 \le t \le 1$. As d grows, this probability quickly approaches 1, even if t is close to 1. This is illustrated in Fig. 1.4.

As shown in Sect. 1.1.3, uniform random points in a d–dimensional ball lie very close to the sphere, the boundary of the ball. Despite this, two-dimensional projections of these points lie far away from the circle, the boundary of the two-dimensional ball; see Fig. 1.5.

Fig. 1.4 If d is large, the radius of a uniformly distributed random point is concentrated very close to 1. Here we plot the probability that this radius exceeds 0.9 as a function of d ($d = 1, \ldots, 50$)

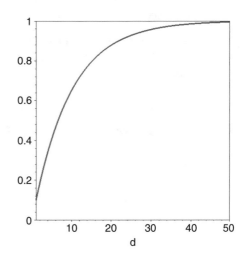

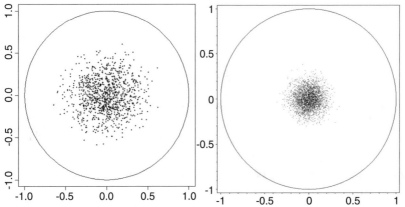

Fig. 1.5 Two-dimensional projections of d-dimensional points uniformly distributed in the unit ball $B_d(1)$. Left: $d = 20$. Right: $d = 50$

Let $d \geq 3$, e be any vector of unit length and $X = (x_1, \ldots, x_d)^\top$ be a uniformly distributed random point inside the unit ball $B_d(1)$. Then, by Theorem 2.7 in [2], for any $c \geq 1$,

$$\Pr\{|e^\top X| \leq c/\sqrt{d-1}\} \geq 1 - \frac{2}{c} e^{-c^2/2}. \tag{1.1.6}$$

The area $\{X : e^\top X = 0\}$ can be called 'equator'. Then (1.1.6) implies that for large d, most of the volume of the ball lies very close to the equator. Moreover, (1.1.6) implies that if X_1 and X_2 are two i.i.d. uniform random vectors on the ball, then X_1 is almost orthogonal to X_2: the angle between X_1 and X_2 is $\pi/2 + O(1/\sqrt{d})$, $d \to \infty$. This statement naturally generalizes to the case of n uniform i.i.d. random vectors; see Theorem 2.8 in [2].

Summary   *In high-dimensional balls, the mass is concentrated close to the boundary and near the equators; uniformly distributed i.i.d. random vectors are almost orthogonal.*

1.1.5   *Distance Between Two Random Points in a Cube*

Assume $X = (x_1, \ldots, x_d)$ and $Y = (y_1, \ldots, y_d)$ are independent random vectors on $C = [0, 1]^d$ consisting of i.i.d. random components x_i and y_i which have some distribution with density $p(x)$, $x \in [0, 1]$. Let $\mu_j = \int_0^1 x^j p(x) dx$ be the moments of the distribution with density $p(x)$. The distribution of the squared distance

$$z = \|X - Y\|^2 = \sum_{i=1}^{d} (x_i - y_i)^2$$

has the mean and variance that can be easily computed as follows:

$$\mathbb{E}z = d\mathbb{E}(x_1 - y_1)^2 = 2d(\mu_2 - \mu_1^2) = 2d\,\mathrm{var}(x_1),$$

$$\mathrm{var}(z) = d\,\mathrm{var}(x_1 - y_1)^2 = d\left[[\mathbb{E}(x_1 - y_1)^4 - [\mathbb{E}(x_1 - y_1)^2)]^2\right]$$

$$= 2d\left[\mu_4 - 4\mu_3\mu_1 + \mu_2^2 + 4\mu_1^2\mu_2 - 2\mu_1^4\right].$$

For example, when x_i and y_i have a Beta(α, α) distribution with density

$$p_\alpha(t) = \frac{1}{\mathrm{Beta}(\alpha, \alpha)}[t(1-t)]^{\alpha-1}, \quad 0 < t < 1, \alpha > 0,$$

where beta$(\cdot, \cdot)$ is the beta-function, then

$$\mu_1 = \frac{1}{2}, \; \mu_2 = \frac{\alpha + 1}{2(2\alpha + 1)}, \; \mu_3 = \frac{\alpha + 2}{4(2\alpha + 1)}, \; \mu_4 = \frac{(\alpha + 2)(\alpha + 3)}{4(2\alpha + 1)(2\alpha + 3)}$$

and therefore

$$\mathbb{E}z = \frac{d}{2(2\alpha + 1)}, \quad \text{var}(z) = d \, \frac{4\alpha + 3}{8 \, (2\,\alpha + 1)^2 \, (2\,\alpha + 3)}. \tag{1.1.7}$$

The expected value of $z = \|X - Y\|^2$ grows proportionally to d (as d increases) and approaches $d/2$ when $\alpha \to 0$. This implies that if we wish the points to be far away from each other, then we have to sample x_i and y_i at 0 w.p. (with probability) 0.5 and at 1 w.p. 0.5. In this case, $\mu_j = 0.5$ for all j and we get $\mathbb{E}z = d/2$ and var$(z) = d/8$. When $\alpha = 1$ (i.e. when X and Y are uniform in the cube C), $\mathbb{E}z = d/6$ and var$(z) = 7d/360 \simeq 0.0194d$.

Summary *(a) The distance between random points grows fast with dimension d, (b) in a sample of n random points $X_1, \ldots, X_n$ in a cube (with a density symmetric around the centre of the cube), all points are at approximately the same distance from each other, and (c) the points are furthest away from each other when the points are sampled (without replacement) from the vertices of the cube.*

1.1.6   Mass Concentration for the Gaussian Distribution

Let $X = (x_1, \ldots, x_d)$ be Gaussian $N(0, I_d)$; then distance from the origin $\|X\|$ is very close to $\sqrt{d}$. This follows from the Central Limit Theorem (CLT) applied to the r.v. $\|X\|^2 = x_1^2 + \ldots + x_d^2$. Moreover, in view of Theorem 2.9 in [2], for any $0 < \beta < \sqrt{d}$,

$$Pr\{\sqrt{d} - \beta \le \|X\| \le \sqrt{d} + \beta\} \ge 1 - 3e^{-\beta^2/96}.$$

Summary *The 'bell-curve' intuition of Fig. 1.6 concerning high-dimensional normal distribution is misleading.*

1.1.7   Volume of a Small Ball Placed In-Between 2^d Large Balls

Consider the cube $C = [-1, 1]^d$ and draw 2^d 'large' balls of radius $\frac{1}{2}$ centred at $(\pm\frac{1}{2}, \ldots, \pm\frac{1}{2})$. Each of these balls touches d other balls. Insert a 'small' ball, the largest ball in the middle of the cube which touches all 'large' balls; see Fig. 1.7. By

Fig. 1.6 Bell-curve:
Gaussian density when $d = 2$

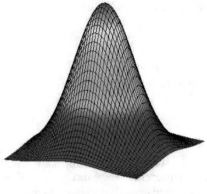

Fig. 1.7 'Small' ball
in-between 2^d large ones,
$d = 2$

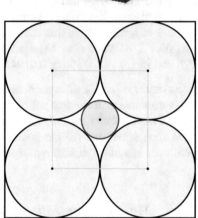

looking at the internal cube $C' = [-1/2, 1/2]^d$, the radius r_d of the 'small' ball is
very easy to compute $r_d = \frac{1}{2}(\sqrt{d} - 1)$.

For example, $r_1 = 0, r_2 \simeq 0.207, r_3 \simeq 0.366, r_4 = \frac{1}{2}, r_9 = 1$ and $r_{100} = 4.5$. For
dimensions $d > 5$, the 'small' ball sticks out of the smaller cube C' and for $d > 10$,
this ball sticks out of the original cube C. For dimensions $d \geq 7$, the volume of the
'small' ball is larger than V_d (defined by (1.1.2)), the total volume of all 2^d 'large'
balls. For dimensions $d \geq 1206$, the volume $r_d^d V_d$ of the 'small' ball is larger than
2^d, the volume of C. Moreover, as $d \to \infty$, the ratio $r_d^d V_d/2^d$ tends to infinity
exponentially fast

$$r_d^d V_d/2^d = \left(\frac{1}{\sqrt{d}} + O\left(\frac{1}{d} \right) \right) \exp \left\{ \frac{1 + \log \pi - 3 \log 2}{2} d - \sqrt{d} - \frac{1}{2} - \log \pi \right\}, \quad d \to \infty,$$

where $\frac{1}{2}(1 + \log \pi - 3 \log 2) \simeq 0.032644$.

Despite the volume of the 'small' ball eventually exceeding the volume of the
cube $C = [-1, 1]^d$, for large d, the volume of intersection of this ball with C is
negligible as this ball is well inside 'the sphere of maximum mass concentration'

discussed in Sect. 1.1.2. For the cube $C = [-1, 1]^d$, this sphere is centred at 0 and has radius $\sqrt{d/3} > r_d = (\sqrt{d} - 1)/2$.

Summary *Two-dimensional geometrical representations of d-dimensional configurations may be very deceiving.*

1.1.8 Volume of Intersection of a Cube and a Ball

The main quantity of interest Consider the following problem. Take the cube $[-1, 1]^d$ of volume $\mathrm{vol}([-1, 1]^d) = 2^d$ and a ball $\mathscr{B}_d(Z, r) = \{Y \in \mathbb{R}^d : \|Y - Z\| \leq r\}$ centred at a point $Z = (z_1, \ldots, z_d)^\top \in \mathbb{R}^d$; this point Z could be outside $[-1, 1]^d$. Denote the fraction of the cube $[-1, 1]^d$ covered by the ball $\mathscr{B}_d(Z, r)$ by

$$C_{d,Z,r} = \mathrm{vol}([-1, 1]^d \cap \mathscr{B}_d(Z, r))/2^d . \tag{1.1.8}$$

As Fig. 1.8 shows, the volume of intersection of a ball and a cube depends on the centre of the ball and could be very different. Our aim is to approximate $C_{d,Z,r}$ for arbitrary d, Z and r. We will derive a CLT-based normal approximation, and then, using an asymptotic expansion in the CLT for non-identically distributed r.v., we will improve it.

Note that in the special case $Z = 0$, several approximations for the quantity $C_{d,0,r}$ have been derived in [40], but their methods cannot be generalized to arbitrary Z.

In Sect. 1.3.4, we will need another quantity which slightly generalizes (1.1.8). Assume that we have a cube $[-\delta, \delta]^d$ of volume $(2\delta)^d$, a ball $\mathscr{B}_d(Z', r') = \{Y \in$

Fig. 1.8 Intersection of two different balls with a cube $d = 2$

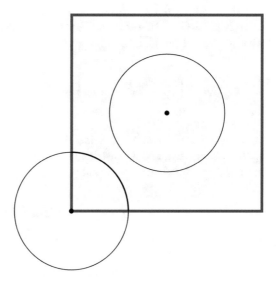

$\mathbb{R}^d : \|Y - Z'\| \leq r'\}$ with a centre at a point $Z' = (z'_1, \ldots, z'_d)^\top$. Denote the fraction of a cube $[-\delta, \delta]^d$ covered by a ball $\mathscr{B}_d(Z', r')$ by

$$C^{(\delta)}_{d,Z',r'} = \text{vol}([-\delta, \delta]^d \cap \mathscr{B}_d(Z', r'))/(2\delta)^d . \tag{1.1.9}$$

By changing the coordinates $Z = Z'/\delta = (z'_1/\delta, \ldots, z'_d/\delta)^\top$ and the radius $r = r'/\delta$, we obtain

$$C^{(\delta)}_{d,Z',r'} = C_{d,Z,r} . \tag{1.1.10}$$

Normal approximation for the quantity (1.1.8) Let $U = (u_1, \ldots, u_d)^\top$ be a random vector with uniform distribution on $[-1, 1]^d$ so that $u_1, \ldots, u_d$ are i.i.d.r.v. uniformly distributed on $[-1, 1]$. Then for given $Z = (z_1, \ldots, z_d)^\top \in \mathbb{R}^d$ and any $r > 0$,

$$C_{d,Z,r} = \mathbb{P}\{\|U - Z\| \leq r\} = \mathbb{P}\left\{\|U - Z\|^2 \leq r^2\right\} = \mathbb{P}\left\{\sum_{j=1}^d (u_j - z_j)^2 \leq r^2\right\} . \tag{1.1.11}$$

That is, $C_{d,Z,r}$, as a function of r, is the c.d.f. of the r.v. $\|U - Z\|$.

Let u have a uniform distribution on $[-1, 1]$ and $z \in \mathbb{R}$. The first three central moments of the r.v. $\eta_z = (u - z)^2$ are

$$\mathbb{E}\eta_z = z^2 + \frac{1}{3}, \quad \text{var}(\eta_z) = \frac{4}{3}\left(z^2 + \frac{1}{15}\right), \quad \mu_z^{(3)} = \frac{16}{15}\left(z^2 + \frac{1}{63}\right), \tag{1.1.12}$$

where $\mu_z^{(3)} = E\,[\eta_z - E\eta_z]^3$.

Consider the r.v. $\|U - Z\|^2 = \sum_{i=1}^d \eta_{z_i} = \sum_{j=1}^d (u_j - z_j)^2$. From (1.1.12), its mean is $\mu_{d,Z} = \mathbb{E}\|U - Z\|^2 = \|Z\|^2 + d/3$. Using independence of $u_1, \ldots, u_d$, we also obtain from (1.1.12) $\sigma_{d,Z}^2 = \text{var}(\|U - Z\|^2) = \frac{4}{3}\left(\|Z\|^2 + \frac{d}{15}\right)$ and

$$\mu_{d,Z}^{(3)} = \mathbb{E}\left[\|U - Z\|^2 - \mu_{d,Z}\right]^3 = \sum_{j=1}^d \mu_{z_j}^{(3)} = \frac{16}{15}\left(\|Z\|^2 + \frac{d}{63}\right) .$$

If d is large enough, then the conditions of the CLT for $\|U - Z\|^2$ are approximately met, and the distribution of $\|U - Z\|^2$ is approximately normal with mean $\mu_{d,Z}$ and variance $\sigma_{d,Z}^2$. That is, we can approximate $C_{d,Z,r}$ by

$$C_{d,Z,r} \cong \Phi\left(\frac{r^2 - \mu_{d,Z}}{\sigma_{d,Z}}\right) , \tag{1.1.13}$$

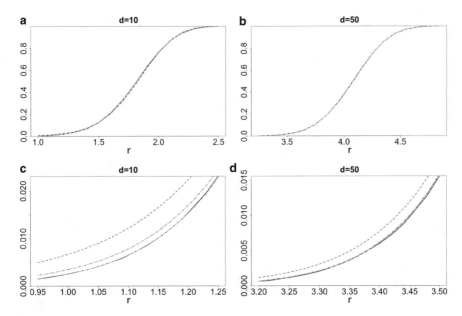

Fig. 1.9 (a) $d = 10$, $Z = 0$, $r \in [1, 2.5]$. (b) $d = 50$, $Z = 0$, $r \in [3.2, 4.9]$. (c) $d = 10$, $Z = 0$, $r \in [0.95, 1.25]$. (d) $d = 50$, $Z = 0$, $r \in [3.2, 3.5]$

where $\Phi(\cdot)$ is the c.d.f. of the standard normal distribution:

$$\Phi(t) = \int_{-\infty}^{t} \phi(v)dv \quad \text{with} \quad \phi(v) = \frac{1}{\sqrt{2\pi}} e^{-v^2/2}.$$

The approximation (1.1.13) has acceptable accuracy if $C_{d,Z,r}$ is not too small; for example, it falls inside a 2σ-confidence interval generated by the standard normal distribution; see Fig. 1.9 as examples. In most interesting cases, this is not true, and the approximation (1.1.13) is not accurate enough and significantly overestimates the true value $C_{d,Z,r}$; see Figs. 1.9–1.11. Below we will improve it by using an Edgeworth-type expansion in the CLT for sums of independent non-identically distributed r.v.

Improved normal approximation General expansion in the CLT for sums of independent non-identical r.v. has been derived by V. Petrov; see Theorem 7 in Chapter 6 in [31] and Sect. 5.6 in [32]. By using only the first term in this expansion, we obtain the following approximation for the distribution function of $\|U - Z\|^2$:

$$P\left(\frac{\|U - Z\|^2 - \mu_{d,Z}}{\sigma_{d,Z}} \leq x\right) \cong \Phi(x) + \frac{\mu_{d,Z}^{(3)}}{6(\sigma_{d,Z}^2)^{3/2}}(1 - x^2)\phi(x),$$

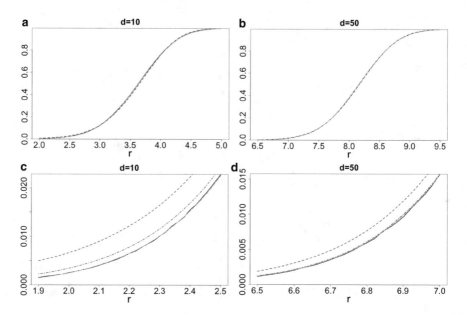

Fig. 1.10 (**a**) $d = 10$, Z is a vertex of $[-1, 1]^d$, $r \in [2, 5]$. (**b**) $d = 50$, Z is a vertex of $[-1, 1]^d$, $r \in [6.5, 9.5]$. (**c**) $d = 10$, Z is a vertex of $[-1, 1]^d$, $r \in [1.9, 2.5]$. (**d**) $d = 50$, Z is a vertex of $[-1, 1]^d$, $r \in [6.5, 7]$

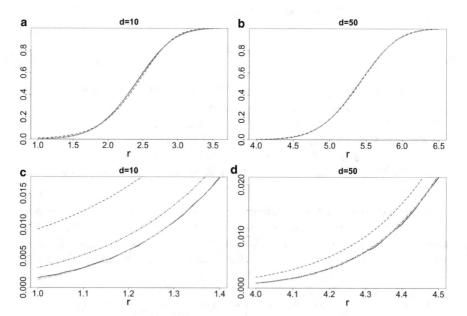

Fig. 1.11 (**a**) Z is at half-diagonal with $\|Z\| = \frac{1}{2}\sqrt{10}$. (**b**) Z is at half-diagonal, $\|Z\| = \frac{1}{2}\sqrt{50}$. (**c**) Z is at half-diagonal, $\|Z\| = \frac{1}{2}\sqrt{10}$. (**d**) Z is at half-diagonal, $\|Z\| = \frac{1}{2}\sqrt{50}$

leading to the following improved form of (1.1.13):

$$C_{d,Z,r} \cong \Phi(t) + \frac{\|Z\|^2 + d/63}{5\sqrt{3}(\|Z\|^2 + d/15)^{3/2}}(1 - t^2)\phi(t), \qquad (1.1.14)$$

where

$$t = t_{d,\|Z\|,r} = \frac{r^2 - \mu_{d,Z}}{\sigma_{d,Z}} = \frac{\sqrt{3}(r^2 - \|Z\|^2 - d/3)}{2\sqrt{\|Z\|^2 + d/15}}. \qquad (1.1.15)$$

The approximation (1.1.14) is much more accurate than the original normal approximation. The other terms in Petrov's expansion of [31] and [32] continue to slightly improve the approximation, but if d is large enough (say, $d \geq 20$), then the approximation (1.1.14) is very accurate, and no further correction is needed.

A very attractive feature of the approximations (1.1.13) and (1.1.15) is their dependence on Z through $\|Z\|$ only. We could have specialized for our case the next terms in Petrov's approximation, but these terms no longer depend on $\|Z\|$ only, and hence the next terms are much more complicated. Moreover, adding one or two extra terms from Petrov's expansion to the approximation (1.1.14) does not fix the problem entirely for all Z and r. Instead, we propose a slight adjustment to the r.h.s of (1.1.14) to improve this approximation, especially for small dimensions. Specifically, the approximation we suggest is

$$C_{d,Z,r} \cong \Phi(t) + c_d \frac{\|Z\|^2 + d/63}{5\sqrt{3}(\|Z\|^2 + d/15)^{3/2}}(1 - t^2)\phi(t), \qquad (1.1.16)$$

where $c_d = 1 + 3/d$ if the point Z lies on the diagonal of the cube $[-1, 1]^d$ and $c_d = 1 + 4/d$ for a typical (random) point Z. For typical (random) points $Z \in [-1, 1]^d$, the values of $C_{d,Z,r}$ are marginally smaller than for the points on the diagonal of $[-1, 1]^d$ having the same norm, but the difference is very small. In addition to the points on the diagonal, there are other special points: the points whose components are all zero except for one. For such points, the values of $C_{d,Z,r}$ are smaller than for typical points Z with the same norm, especially for small r. Such points, however, are of no value for us as they are not typical and we have never observed in simulations random points that come close to these truly exceptional points.

Simulation study In Figs. 1.9–1.11, we demonstrate the accuracy of approximations (1.1.13), (1.1.14) and (1.1.16) for $C_{d,Z,r}$ in dimensions $d = 10, 50$ for the following locations of Z: (i) $Z = 0$, the centre of the cube $[-1, 1]^d$; (ii) $\|Z\| = \sqrt{d}$, with Z being a vertex of the cube $[-1, 1]^d$; (iii) Z lies on a diagonal of $[-1, 1]^d$ with $|z_j| = \lambda \geq 0$ for all $j = 1, \ldots, d$ and $\|Z\| = \lambda\sqrt{d}$.

There are figures of two types. In the figures of the first type, we plot $C_{d,Z,r}$ over a wide range of r ensuring that values of $C_{d,Z,r}$ lie in the whole range $[0, 1]$. In the figures of the second type, we plot $C_{d,Z,r}$ over a much smaller range of r with $C_{d,Z,r}$ lying in the range $[0, \varepsilon]$ for some small positive ε such as $\varepsilon = 0.015$. As we need to assess the accuracy of all approximations for smaller values of $C_{d,Z,r}$, the second type of plots are often more insightful. In Figs. 1.9–1.11, the solid black line depicts values of $C_{d,Z,r}$ computed via Monte Carlo methods; the blue-dashed, the red dot-dashed and green long-dashed lines display approximations (1.1.13), (1.1.14) and (1.1.16), respectively.

From the simulations that led to Figs. 1.9–1.11, we can make the following conclusions: (a) the normal approximation (1.1.13) is quite satisfactory unless the value $C_{d,Z,r}$ is small; (b) the accuracy of all approximations improves as d grows; (c) the approximation (1.1.16) is very accurate even if the values $C_{d,Z,r}$ are very small; and (d) if d is large enough, then the approximations (1.1.14) and (1.1.16) are practically identical and are extremely accurate.

Summary   *Using an asymptotic expansion in the CLT, we are able to accurately approximate the proportion of a d-dimensional cube covered by a ball centred at any point in $\mathbb{R}^d$.*

1.2   Space-Filling, Uniformity and Discrepancies

In this section, we will view space-filling from the very common viewpoint of uniformity characterized via one of the discrepancies. Also, in Sect. 1.2.5, we briefly consider some other characteristics of space-filling.

Let the set to be filled be $\mathscr{X}$, a bounded set in $\mathbb{R}^d$, with the main example $\mathscr{X} = [0, 1]^d$. An n-point design $\mathbb{Z}_n$ is a collection of n points in $\mathscr{X}$: $\mathbb{Z}_n = \{z_1, \ldots, z_n\} \subset \mathscr{X}$. The following two cases are often distinguished: (i) n is fixed (non-nested designs), and (ii) $z_1, \ldots, z_n$ are the first n points from the sequence $\mathbb{Z}_\infty = \{z_1, z_2, \ldots\}$ (nested designs).

If we think about a space-filling design as a design resembling the one from Fig. 1.12, then it is natural to associate space-filling with uniformity and hence

Fig. 1.12 A space-filling design, $d = 2$

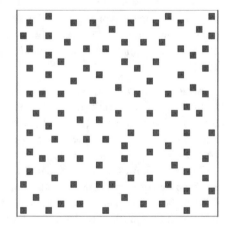

characterize the quality of space-filling by one of the discrepancies discussed below. Note, however, that in view of numerous discussions in Sect. 1.1, two-dimensional representations of high-dimensional sets and designs could be misleading.

1.2.1   Discrepancies

Discrepancies quantify the uniformity of n-point sets and characterize the deviation between the empirical probability measure μ_n, which assigns weights $1/n$ to each point of the design $\mathbb{Z}_n = \{z_1, \ldots, z_n\}$, and μ, the uniform measure on $\mathscr{X}$.

There are various different discrepancies considered in literature (see, e.g. [9, 16, 25, 26]). We only discuss several of them assuming additionally $\mathscr{X} = [0, 1]^d$:

- local discrepancy: $D_n(\alpha) = \left| \frac{1}{n} \sum_{i=1}^{n} 1_{\{z_i \leq \alpha\}} - \alpha_1 \cdots \alpha_d \right|, \alpha = (\alpha_1, \cdots, \alpha_d) \in \mathscr{X}$;
- $D_{n,f} = \left| \frac{1}{n} \sum_{i=1}^{n} f(z_i) - \int_{\mathscr{X}} f(x)dx \right|$, where f is a measurable function on $\mathscr{X}$;
- star-discrepancy: $D_n^* = \max_{\alpha \in \mathscr{X}} D_n(\alpha)$;
- L_p-discrepancy: $D_n^{(p)} = \left[\int_x D_n^p(\alpha)d\alpha \right]^{1/p}$;
- kernel (maximum-mean) discrepancy $\gamma_n(\mu_n, \mu) = \left[\iint K(x, y)(\mu_n - \mu)(dx) (\mu_n - \mu)(dy) \right]^{1/2}$, where $K(\cdot, \cdot)$ is a kernel;
- $\varrho(\mu_n, \mu)$, where ϱ is any distance between probability measures;
- weighted discrepancies, where points z_i are assigned weights.

Star-discrepancy and its importance   The usual definition of a uniformly distributed sequence is stated in terms of star-discrepancy as follows.

Definition   A sequence $\mathbb{Z}_\infty = \{z_1, z_2, \ldots\} \subset [0, 1]^d$ is uniformly distributed on $[0, 1]^d$ if $D_n^* \to 0$ as $n \to \infty$.

The importance of the star-discrepancy D_n^* is mostly related to the so-called Koksma-Hlawka inequality: for any design $\mathbb{Z}_n$,

$$\left| \frac{1}{n} \sum_{i=1}^{n} f(z_i) - \int_{\mathscr{X}} f(x)dx \right| \leq V(f)D_n^*, \tag{1.2.1}$$

where $V(f)$ is the variation of f (in the sense of Hardy and Krause) on $\mathscr{X} = [0, 1]^d$. Moreover, there exist functions f such that $V(f) = 1$, and for all $\varepsilon > 0$, we have $\left| \frac{1}{n} \sum_{i=1}^{n} f(z_i) - \int_{\mathscr{X}} f(x)dx \right| > (D_n^* - \varepsilon)$. Note, however, that for large d, the variation (in the sense of Hardy and Krause) of a typical function f can be astronomically large and hence the upper bound in (1.2.1) has little practical sense if d is not small.

Another reason for popularity of the star-discrepancy D_n^* is the fact that one of the main conjectures in numerical mathematics concerns the behaviour of D_n^* for large n. This conjecture has two equivalent forms:

$$\exists\, c > 0 : \limsup_n \left[n D_n^* / \log^d n \right] \geq c \ \text{ for any } \mathbb{Z}_\infty = \{z_1, z_2, \ldots\}, \qquad (1.2.2)$$

$$\exists\, c > 0 : \limsup_n \left[n D_n^* / \log^{d-1} n \right] \geq c \ \text{ for any family of designs } \mathbb{Z}_n. \quad (1.2.3)$$

The conjecture (1.2.2) has led to the important concept of a low-discrepancy sequence considered in the next section. We shall come back to the star-discrepancy in Sect. 1.2.4.

Low-discrepancy sequences.

Definition If for a sequence $\mathbb{Z}_\infty$, $D_n^* \leq C \log^d n / n$, for some $C > 0$ and all n, then $\mathbb{Z}_\infty$ is called a low-discrepancy sequence.

Similar definition holds for families of non-nested designs, where the inequality takes the form $D_n^* \leq C \log^{d-1} n / n$.

Many low-discrepancy sequences (and point sets) are known. Some of them are very popular. There is, however, the following widespread illusion:

An illusion *any low-discrepancy sequence (or a low-discrepancy family of non-nested designs) has good space-filling properties.*

This is a very important point we want to make in this chapter, and we shall return to this issue several times. For the moment, we will consider a very simple example of a lattice like the one depicted in Fig. 1.13. Formally, lattices possess many wonderful properties. However, there are at least two problems with lattices (as with all other classical low-discrepancy sequences): (a) good space-filling properties are observed only for very special values of n, and, even more importantly, (b) for large d, one has to choose astronomically large values of n to be able to see any space-filling delivered by such sequences.

For the lattice in Fig. 1.13, we observe good space-filling only if either we see fat points only or when we finish placing thin points. For large d, we are never going to see any good space-filling properties of any lattice as almost all the volume of the

Fig. 1.13 A lattice

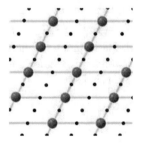

cube $[0, 1]^d$ is near its boundary (see Sect. 1.1.3), but the space-filling properties of lattices are observed in the interior of the cube only.

Different rates for different discrepancies as n increases If z_i are uniform i.i.d. on $[0, 1]^d$, then the star-discrepancy and all L_p-discrepancies have rate of decrease $n^{-1/2}$, $n \to \infty$, $\forall d$ (with probability $1 - \gamma$, $\forall \gamma > 0$). However, if z_i are deterministic points, then different discrepancies may have different rates of decrease as $n \to \infty$. Moreover, studying these rates for different discrepancies may lead to problems of completely different complexities. Let us give two examples:

(i) In contrast to conjecture (1.2.2), a similar lower bound for the L_p-discrepancy is $\limsup_n D_n^{(p)} \geq c_{d,p} \log^{d/2} n/n$, and it is not hard to prove that this lower bound is achievable for certain low-discrepancy sequences.

(ii) The Farey sequence $\mathscr{F}_Q$ of order Q consists of the rational numbers (Farey fractions) $a/q \in [0, 1]$ with $1 \leq q \leq Q$ such that $0 \leq a/q \leq 1$; note $0 = 0/1, 1 = 1/1 \in \mathscr{F}_Q$. The total number of fractions in $\mathscr{F}_Q$ is $n = F_Q + 1 = \sum_{q=1}^{Q} \varphi(q) + 1$, where $\varphi(\cdot)$ is the Euler's totient function. As we do not count the repeats again, the highest common factor of a and q must be one; for each $q > 1$ there are $\varphi(q)$ possible values of a. It is well-known that $F_Q = 3Q^2/\pi^2 + O(Q \log Q)$ as $Q \to \infty$. The asymptotic distribution (as $Q \to \infty$) of the spacings for the Farey sequences $\mathscr{F}_Q$ is well-studied; see, e.g. [17, 22]. The results use the following fundamental property of $\mathscr{F}_Q$: the spacings $p_i = x_{(i+1)} - x_{(i)}$ in $\mathscr{F}_Q$ are $1/(qq')$ for $x_{(i)} = a/q$, $x_{(i+1)} = a'/q'$. From [8], for all $Q > 1$, $D_n^* = 1/Q$. The classical results of Franel [12] and Landau [23] imply that the celebrated Riemann hypothesis (RH) is equivalent to the statement that the decrease of the L_2-discrepancy of the Farey sequence is of order $O\left(Q^{-3/2+\varepsilon}\right)$, $\forall \varepsilon > 0$, as $Q \to \infty$; this is a faster decrease than $D_n^* = 1/Q$. Moreover, in a series of recent papers, Kanemitsu and Yoshimoto (see, e.g. [20, 21]) have shown that the RH is also equivalent to the statement that the local discrepancy $D_n(\alpha)$ at any fixed rational $\alpha \neq \frac{1}{2}$ has the same order $O\left(Q^{-3/2+\varepsilon}\right)$, $\forall \varepsilon > 0$ (here a version of GRH, generalized Riemann hypothesis, is also assumed). An illustration of this result is given in Fig. 1.14, where we plot the normalized discrepancies $Q^{3/2} D_n(\alpha)$ for several rational α. Under the RH, these graphs should stay below CQ^ε for some C and all $\varepsilon > 0$. Periodograms of the series of Fig. 1.14 are depicted in Fig. 1.15.

1.2.2 Weighted Discrepancies

Regular discrepancies characterize the deviation

$$D_{n,f} = \left| \frac{1}{n} \sum_{i=1}^{n} f(z_i) - \int_{\mathscr{X}} f(x) dx \right|$$

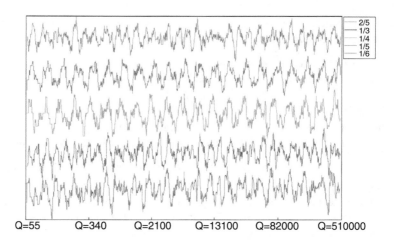

Fig. 1.14 Discrepancies of the Farey series at rational points as functions of log Q

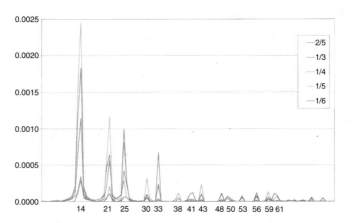

Fig. 1.15 Periodograms of the series of Fig. 1.14 as functions of $\gamma = 2\pi \times$ frequency

for various classes of functions f. For example, the star-discrepancy D_n^* is the supremum of $D_{n,f}$ over the class of indicator functions $f_\alpha(x) = 1_{[x \leq \alpha]}$ with $\alpha \in [0, 1]^d$. Weighted discrepancies measure the deviation

$$D_{n,f}^{(W)} = \left| \sum_{i=1}^{n} w_i f(z_i) - \int_{\mathscr{X}} f(x) dx \right|$$

where w_i are weights assigned to z_i. The usual assumption for the weights is $\sum_{i=1}^{n} w_i = 1$ (but the weights do not have to be non-negative). Weighted star-discrepancy for $\mathscr{X} = [0, 1]^d$ is then defined as $D_n^{*(W)} = \max_{\alpha \in \mathscr{X}} D_n^{(W)}(\alpha)$, where

Fig. 1.16 Voronoi cells

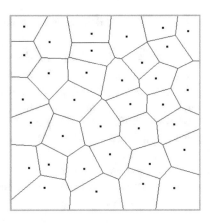

$$D_n^{(W)}(\alpha) = \left| \sum_{i:z_i \leq \alpha}^{n} w_i - \alpha_1 \cdots \alpha_d \right| ;$$

here $u \leq v$ for $u, v \in \mathbb{R}^d$ if all components of u are smaller or equal than the corresponding components of v.

It is shown in [38, Th.1.2, Ch.2] that if we define the weighted discrepancy by $D_n^{(W)}(\text{Lip}_L) = \sup_{f \in \mathscr{F}} D_{n,f}^{(W)}$, where $\mathscr{F}$ is the class of Lipschitz functions on $\mathscr{X}$ with Lipschitz constant $L > 0$, then for any nodes $z_1, \ldots, z_n$, the optimal weights w_i are proportional to the volumes of the Voronoi cells (the result does not depend on the value of the Lipschitz constant L). In the case $\mathscr{X} = [0, 1]$, a similar result holds for the weighted star-discrepancy; see Sect. 1.3.3. The Voronoi cell V_i associated with the point z_i is $V_i = \{x \in \mathscr{X} : \|x - z_i\| \leq \|x - z_j\|, j \neq i\}$, see Fig. 1.16. Voronoi cells can be thought of as the regions of attraction for the objective function $f(x) = \min_{i=1,\ldots,n} \|x - z_i\|^2$.

In view of the conjectured lower bound (1.2.2), we can formulate the following conjecture.

Conjecture *Let* $\mathscr{X} = [0, 1]^d$ *and* $\{w_1, \ldots, w_n\}$ *be the optimal set of weights minimizing the weighted star-discrepancy* $D_n^{*(W)}$ *for given* $\mathbb{Z}_n = \{z_1, \ldots, z_n\}$. *Then: (a) there exists* $c > 0$ *such that* $\limsup_n n D_n^{*(W)}/\log^{d-1} n \geq c$ *for any sequence* $\mathbb{Z}_\infty = \{z_1, z_2, \ldots\} \subset \mathscr{X}$, *and (b) there exists a sequence* $\mathbb{Z}_\infty = \{z_1, z_2, \ldots\}$ *with* $\limsup_n n D_n^{*(W)}/\log^{d-1} n < \infty$.

Similar conjectures can be formulated for some other weighted discrepancies, in particular for $D_n^{(W)}(\text{Lip}_L)$. The rationale behind this conjecture is that the assignment of weights allows us to reduce the intrinsic dimensionality, from d to $d - 1$.

1.2.3   Weighted Discrepancies for Farey Sequences and the RH

Let us continue the discussion started at the end of Sect. 1.2.1 and consider the weighted Farey sequences $\mathscr{F}_Q$, where the fractions a/q are assigned some weights $w(q, Q)$. The 'most uniform' weighted Farey sequence is when $w(q, Q) = 1/q$; for large Q, this weighting is equivalent to counting the repeats of a/q as many times as they occur. In this case, all the weighted discrepancies defined above, including the weighted star-discrepancy, have the order $O\left(Q^{-2}\right)$ as $Q \to \infty$ and there is no relation to the RH.

There are, however, many other rules for assigning the weights where (a) we can clearly observe the correspondence between the RH, weighted local discrepancies at rational points and weighted L_2 discrepancy and (b) the behaviour of weighted discrepancies is much smoother than the behaviour of similar discrepancies for the original Farey sequences. For illustration, in Fig. 1.17 we plot the normalized weighted discrepancies $Q^{3/2} D_n^{(W)}(\alpha)$ for the same values of α as in Fig. 1.14; the weights chosen for Fig. 1.17 are $w(q, Q) = (1 - q/Q)^2$. Periodograms of the series of Fig. 1.17 (not shown), similar to the one of Fig. 1.15, also allow to numerically identify several few zeros of $\zeta(\cdot)$. By comparing Figs. 1.14 and 1.17, we clearly see that the series of Fig. 1.17 are much smoother. These series may well have a simpler structure and even reveal new ways of proving the RH.

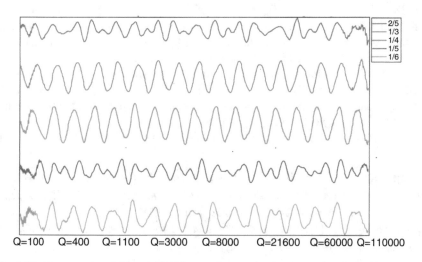

Fig. 1.17 Discrepancies of the weighted Farey series at rational points as functions of $\log Q$

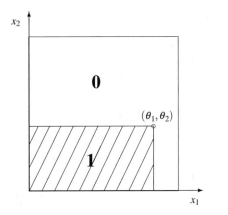 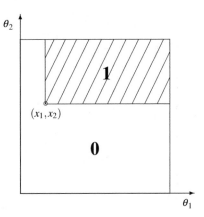

Fig. 1.18 Left: Indicator function $\mathbf{1}_{[0,\theta)}(x)$. Right: Partition of $\Theta = [0, 1)^d$ into two consistent sets

1.2.4   Star-Discrepancy as Integration Error for Indicator Functions

As follows from the definition, the star-discrepancy D_n^* is $D_n^* = \sup_{\theta \in \Theta} |\hat{I}_\theta - I_\theta|$, the maximum estimation error of the integral $I_\theta = \int_{[0,1]^d} f(x, \theta)dx = \prod_{i=1}^d \theta_i$ of the indicator functions (illustrated at Fig. 1.18, left)

$$f(x, \theta) = \mathbf{1}_{[0,\theta)}(x) = \begin{cases} 0 \text{ if } x \le \theta \\ 1 \text{ if } x > \theta \end{cases}$$

by $\hat{I}_\theta = \frac{1}{n} \sum_{i=1}^n f(x_i, \theta)$, where $\theta = (\theta_1, \ldots, \theta_d) \in \Theta$ and $\Theta = [0, 1)^d$. In view of (1.2.3), for any design $\mathbb{Z}_n$, D_n^* has the order of $\log^{d-1}(n)/n$ as $n \to \infty$.

In this section, we show that the integration error may achieve the order $O(1/n)$, $n \to \infty$, by means of replacement of $\hat{I}_\theta$ with some other estimates. We also construct the designs which achieve the minimal integration error in a class of product designs.

Evaluation of the indicator function leads to the partition of $\Theta = [0, 1)^d$ into two sets (which we call consistent sets). These are the subsets of Θ leading to the same evaluation results; see Fig. 1.18 (right). This can be considered as dual to Fig. 1.18 (left). Figure 1.19 (left) demonstrates a typical partition of $\Theta = [0, 1)^d$ produced by several observations of the indicator function, and Fig. 1.19 (right) shows construction of the design which generates the partition with the maximum number of consistent sets.

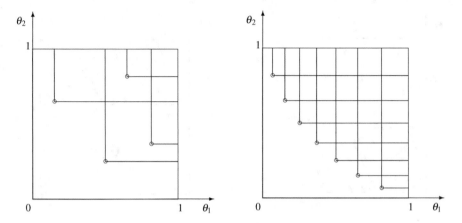

Fig. 1.19 Left: Typical partition of Θ into consistent sets. Right: Partition of Θ with the maximum number of consistent sets

Consider the family of product designs $\mathbb{Z}_n$ consisting of $n = dk$ points of the form

$$
\begin{aligned}
(z_i, 0, 0, \ldots) \ & (i = 1, \ldots, k) \\
(0, z_i, 0, \ldots) \ & (i = 1, \ldots, k) \\
\cdots \qquad\quad & \cdots \\
(0, 0, \ldots z_i) \ & (i = 1, \ldots, k)
\end{aligned}
\tag{1.2.4}
$$

where $0 = z_0 < z_1 < \cdots < z_k < 1 = z_{k+1}$.

As illustrated by Fig. 1.20, in this case the observations partition the parameter space $\Theta = [0, 1)^d$ into rectangular consistent sets

$$
\Theta(i_1, \ldots, i_d) = \otimes_{j=1}^{d} [z_{i_j}, z_{i_j+1})
\tag{1.2.5}
$$

where $i_1, \ldots, i_d$ are integers, $0 \le i_j \le k$ for all $j = 1, \ldots, d$.

The minimal and maximal values for the volume $I_\theta = \prod_{j=1}^{d} \theta_j$ as $\theta = (\theta_1, \ldots, \theta_d)$ ranges over (1.2.5) are equal to $\prod_{j=1}^{d} z_{i_j}$ and $\prod_{j=1}^{d} z_{i_j+1}$, respectively. Estimating I_θ for $\theta \in \Theta(i_1, \ldots, i_d)$ by the worst-case optimal estimator

$$
\tilde{I}_\theta = \frac{1}{2} \left(\prod_{j=1}^{d} z_{i_j} + \prod_{j=1}^{d} z_{i_j+1} \right)
\tag{1.2.6}
$$

we obtain the following expression for the maximal error of estimate $\tilde{I}_\theta$ of I_θ for $\theta \in \Theta(i_1, \ldots, i_d)$:

$$
V(i_1, \ldots, i_d) = \sup_{\theta \in \Theta(i_1, \ldots, i_d)} |\tilde{I}_\theta - I_\theta| = \frac{1}{2} \left(\prod_{j=1}^{d} z_{i_j+1} - \prod_{j=1}^{d} z_{i_j} \right).
$$

Fig. 1.20 Consistent sets for the design (1.2.4), $d = 2, k = 7$

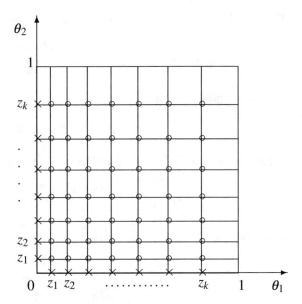

As a design criterion, consider the worst-case criterion

$$V_{WC}(\mathbb{Z}_n) = V_{WC}(z_1, \ldots, z_k) = \sup_{\theta} |\tilde{I}_\theta - I_\theta| =$$

$$\sup_{i_1,\ldots,i_s} \sup_{\theta \in \Theta(i_1,\ldots,i_d)} |\tilde{I}_\theta - I_\theta| = \frac{1}{2} \sup_{i_1,\ldots,i_d} \left(\prod_{j=1}^{d} z_{i_j+1} - \prod_{j=1}^{d} z_{i_j} \right).$$

Equally spaced points $z_i = i/(k+1)$, $i = 1, \ldots, k$, give the hypercube $[\frac{k}{k+1}, 1)^d$ as the hyper-rectangle with the largest value of error $|\tilde{I}_\theta - I_\theta|$ and therefore

$$V_{WC}\left(\frac{1}{k+1}, \ldots, \frac{k}{k+1} \right) = \frac{1}{2} \left(1 - \left(\frac{k}{k+1} \right)^d \right) \sim \frac{d}{2k} = \frac{d^2}{2n} \quad \text{as } n \to \infty.$$

This simple choice of design points already demonstrates that the rate of convergence is at least const/n, $n \to \infty$. Theorem 1.1 and the fact that for any n-point design the number of consistent sets is at most $(n^d/d!)(1 + o(1))$, $n \to \infty$, (see Fig. 1.19, right) yield that const/n, $n \to \infty$, is the best possible worst-case rate. The following theorem solves the optimum design problem in the class of product designs (1.2.4) for the criterion $V_{WC}(z_1, \ldots, z_k)$ and establishes the asymptotical expression for $\min_{z_i} V_{WC}(z_1, \ldots, z_k)$, when $k \to \infty$.

Theorem 1.1 *Within the class of the product designs* (1.2.4) *with* $n = dk$

$$\min_{z_1,\ldots,z_k} V_{WC}(z_1,\ldots,z_k) = 1 - a^d \sim \frac{d^2 \log d}{2(d-1)} \frac{1}{n} \quad \text{as } n \to \infty, \quad (1.2.7)$$

where a satisfies the equation

$$a^{k(d-1)}(a^d - 1) - a + 1 = 0, \quad 0 < a < 1, \quad (1.2.8)$$

and the minimum in (1.2.7) *is achieved for the following choice of* z_i: $z_k = a$,

$$z_m = (1 - a^d)\frac{1 - a^{m(d-1)}}{1 - a^{d-1}} \quad \text{for } m = 1,\ldots,k-1. \quad (1.2.9)$$

Proof Let $n = dk$ and consider the product design (1.2.4). Then

$$V_{WC}(z_1,\ldots,z_k) = \frac{1}{2} \sup_{i_1,\ldots,i_d} e(i_1,\ldots,i_d)$$

where

$$e(i_1,\ldots,i_d) = \prod_{j=1}^{d} z_{i_j+1} - \prod_{j=1}^{d} z_{i_j}.$$

The statement that $e(i_1,\ldots,i_d)$ all equal when $i_1 = i_2 =,\ldots,= i_{d-1}$ and $i_d \in \{1,\ldots,k\}$ leads to the equations

$$z_1 = z_2 - z_1 z_k^{s-1} = \ldots = z_m - z_{m-1} z_k^{s-1} = \ldots = 1 - z_k^d. \quad (1.2.10)$$

Now define $a = z_k$, $c = 1 - z_k^d$, $b = a^{d-1}$; then the solution to (1.2.10) is given by

$$z_1 = c, \quad z_2 = c(1+b), \ldots, z_m = c\frac{1-b^m}{1-b}, \ldots, z_k = c\frac{1-b^k}{1-b} = a.$$

This yields the equation (1.2.8) for a and the expressions (1.2.9) for z_m.

We now show that in the case of equal values of $e(i_1,\ldots,i_d)$ on the edge hyper-rectangles

$$\Theta(i_1,\ldots,i_d) = \otimes_{j=1}^{d}[z_{i_j}, z_{i_j+1})$$

with $i_j = k$ for at least $d-1$ values of $j \in \{1,\ldots,d\}$, the values of $e(i_1,\ldots,i_d)$ are smaller for the internal hyper-rectangles. For this, it is enough to demonstrate that the function $e(i_1,\ldots,i_d)$ is pointwise increasing in any index. Without loss of generality, we show this for the first index. Thus, for $i_1 < k$

$$e(i_1 + 1, \ldots, i_d) - e(i_1, \ldots, i_d) = (z_{i_1+2} - z_{i_1+1})z_{i_2+1}, \ldots, z_{i_d+1} -$$

$$(z_{i_1+1} - z_{i_1})z_{i_2}, \ldots, z_{i_d} = cb^{i_1+1}z_{i_2+1}, \ldots, z_{i_d+1} - cb^{i_1}z_{i_2}, \ldots, z_{i_d} =$$

$$cb^{i_1}(bz_{i_2+1}, \ldots, z_{i_d+1} - z_{i_2}, \ldots, z_{i_d}).$$

We need to show that this is non-negative. It will hold if $b^{1/(d-1)}z_{m+1} \geq z_m$ or, equivalently, $az_{m+1} \geq z_m$ for all $m = 1, \ldots, k$. Converting this statement to a statement in a, we have $a\left(1 - a^{(m+1)(d-1)}\right) \geq 1 - a^{m(d-1)}$ or

$$a^{(m+1)(d-1)} - a^{m(d-1)} - a + 1 \leq 0. \tag{1.2.11}$$

This is the equation (1.2.8) with k replaced with m and equality with inequality. Since the left hand side of (1.2.11) is monotonously increasing in m and $m \leq k$, the inequality (1.2.11) holds.

To complete the proof, we only need to establish the asymptotics in (1.2.7). After removal of the factor $(a - 1)$, the equation (1.2.8) becomes $a^{k(d-1)}(1 + a + \ldots + a^{d-1}) - 1 = 0$ which leads to the following bounds for a:

$$d^{-1/(kd-k)} \leq a \leq d^{-1/(kd-k+d-1)}.$$

This implies, when $k \to \infty$, $a^d \sim (1d)^{-d/(kd-k)}$ when $k \to \infty$, which in its turn yields the asymptotic relation in (1.2.7). ∎

The points $z_1, \ldots, z_k$ may typically be defined as $z_i = G(i/(k+1))$, $i = 1, \ldots, k$ where G is some c.d.f. If G corresponds to an absolutely continuous distribution, then, when $k \to \infty$, the points z_i have the asymptotic density $\phi(z)$ which relates to G through $\phi(z) = dG^{-1}(z)/dz$.

Equally spaced points $z_i = i/(k+1)$, $i = 1, \ldots, k$, correspond to $G(z) = z$ and the uniform density on $[0, 1]$. The points z_i of Theorem 1.1 have the asymptotic density

$$\phi_*(z) = dG_*^{-1}(z)/dz = \frac{1}{\log d}\frac{1}{\frac{d}{d-1} - z}, \quad 0 \leq z \leq 1;$$

the corresponding c.d.f. is $G_*(t) = \frac{d}{d-1}(1 - d^{-t})$, $0 \leq t \leq 1$.

1.2.5   Other Characteristics of Space-Filling

The following three quantities are the most common characteristics of space-filling.

Covering radius   *The covering radius* CR *of* $\mathbb{Z}_n = \{z_1, \ldots, z_n\}$ *is defined by*

$$\mathrm{CR}(\mathbb{Z}_n) = \max_{x \in \mathcal{X}}\ \min_{i=1,\ldots,n}\ \|x - z_i\|. \tag{1.2.12}$$

Quantization error *For a given set of points* $\mathbb{Z}_n = \{z_1, \ldots, z_n\} \subset \mathcal{X}$, *the mean squared quantization error is*

$$\text{QE}(\mathbb{Z}_n) = \frac{1}{\text{vol}(\mathcal{X})} \int_{\mathcal{X}} \min_{z_i \in \mathbb{Z}_n} \|x - z_i\|^2 dx = \frac{1}{\text{vol}(\mathcal{X})} \sum_{i=1}^{n} \int_{V_i} \|x - z_i\|^2 dx, \quad (1.2.13)$$

where V_i is the Voronoi set associated with z_i; see Fig. 1.16.

Packing radius *The packing radius* (PR) *of the set of points* $\mathbb{Z}_n = \{z_1, \ldots, z_n\}$ *is defined by*

$$\text{PR}(\mathbb{Z}_n) = \frac{1}{2} \min_{i \neq j} \|z_i - z_j\|. \quad (1.2.14)$$

Another common name for packing radius is separation radius. Also, a popular in computer experiments maximin-distance criterion is $2\,\text{PR}(\mathbb{Z}_n)$. Note that unlike in the case of packing considered in http://packomania.com and some other (mostly, recreational) literature, parts of the balls with centres at z_i are allowed to be outside $\mathcal{X}$. We also point out that packing in the sense of definition (1.2.14) is very different from ball packing in the whole space $\mathbb{R}^d$, which is a very elaborate mathematical subject; see, e.g. [6] and [14, Ch.30]. Unlike the problem of packing balls in a given set leading to (1.2.14), there are no boundary effects in the problem of ball packing in the whole space.

CR is a much more natural characteristic of space-filling than PR (we postpone the discussion of covering until Sect. 1.3). However, for a given set of points $\mathbb{Z}_n$, PR is usually much easier to compute than the CR, and this is the main reason explaining widespread use of PR as a space-filling characteristic. There is sometimes a hope that $\mathbb{Z}_n$ with large PR may have small CR. Indeed, the 'mesh ratio' (or 'mesh-separation ratio') $\text{CR}(\mathbb{Z}_n)/\text{PR}(\mathbb{Z}_n)$ is ≥ 0.5 if $\mathcal{X}$ is connected and $\text{CR}(\mathbb{Z}_n)/\text{PR}(\mathbb{Z}_n) \leq 1$ for (some) sets of points $\mathbb{Z}_n$ with largest packing radius (these point sets have minimal number of pairs of points z_i, z_j such that $\|z_i - z_j\| = 2\,\text{PR}(\mathbb{Z}_n)$).

Consider now a counterexample to a plausible suggestion that large PR implies small CR and vice versa. Consider the two-dimensional unit ball $\{(x, y)^{\top} : x^2 + y^2 \leq 1\}$ and two five-point sets

$$X_5^{(1)} = \left\{ \begin{pmatrix} \cos(2\pi k/5) \\ \sin(2\pi k/5) \end{pmatrix} \right\}_{k=0}^{4} \quad \text{and} \quad X_5^{(2)} = \left\{ \begin{pmatrix} \lambda \cos(2\pi k/5) \\ \lambda \sin(2\pi k/5) \end{pmatrix} \right\}_{k=0}^{4}.$$

where $\lambda = (\sqrt{5} - 1)/2 \simeq 0.618034$.

We have PR $(X_5^{(1)}) = \sqrt{10 - 2\sqrt{5}}/4 \simeq 0.587785$ and $X_5^{(1)}$ is clearly PR - optimal five-point set. For $X_5^{(2)}$, we have PR $(X_5^{(2)}) = \sqrt{5 - 2\sqrt{5}}/2 \simeq 0.36327127$, so $X_5^{(2)}$ is poor with respect to packing. However, $X_5^{(2)}$ is CR-optimal with $\text{CR}(X_5^{(2)}) = \lambda$. The covering radius of $X_5^{(1)}$ is 1, which is very poor. This example is illustrated in Fig. 1.21.

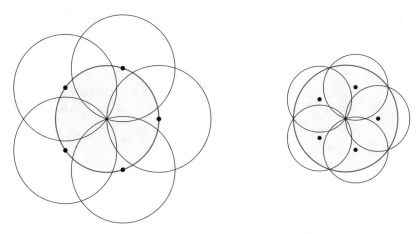

Fig. 1.21 Coverings produced by the two five-point sets. Left: $X_5^{(1)}$. Right: $X_5^{(2)}$. The set $X_5^{(1)}$ is PR -optimal with PR $(X_5^{(1)}) \simeq 0.5878$, and $X_5^{(2)}$ is CR-optimal with CR$(X_5^{(2)}) \simeq 0.618$. At the same time, $X^{(1)}$ produces very bad covering with CR$(X_5^{(1)}) = 1$, and $X_5^{(2)}$ generates bad packing with PR $(X_5^{(2)}) \simeq 0.363$

PR$(\mathbb{Z}_n)$ is easy to compute but difficult to maximize with respect to the location of points z_i as the criterion PR$(\mathbb{Z}_n)$ is non-differentiable with respect to z_i's. Similarly to approximation of max, which defines the L_∞-norm, with the L_q-norm $\| \cdot \|_q$ with large q, it is common to approximate $\min |f|$ with $1/\|1/f\|_q$ with large q. In this way, we obtain the following regularized version PR$^{(q)}(\mathbb{Z}_n)$ of PR$(\mathbb{Z}_n)$:

$$\mathsf{PR}^{(q)}(\mathbb{Z}_n) = \left[\frac{2}{n(n-1)} \sum_{1 \leq i < j \leq n} \|z_i - z_j\|^{-q} \right]^{-1/q} , \quad q > 0. \quad (1.2.15)$$

Maximization of PR$^{(q)}(\mathbb{Z}_n)$ with respect to $z_1, \ldots, z_n$ is equivalent to finding n Fekete points z_i that minimize the q-energy $\sum_{1 \leq i < j \leq n} \|z_i - z_j\|^{-q}$. Continuous version of this energy is the functional

$$E_q(\zeta) = \int_{\mathscr{X}} \int_{\mathscr{X}} \|z - t\|^{-q} \, \zeta(\mathrm{d}t) \, \zeta(\mathrm{d}z), \quad 0 < q < d.$$

The determination of the minimum-energy probability measures, which are called q-equilibrium measures, is one of the main topics in potential theory.

Additionally to the three space-filling characteristics defined above (CR, PR and QE), the following characteristic can also be of interest.

Spacing radius *For a given set of points $\mathbb{Z}_n$, the spacing radius* SR$(\mathbb{Z}_n)$ *is defined as the radius of the largest open ball which fully belongs to $\mathscr{X}$ and does not contain points $z_i \in \mathbb{Z}_n$; see Fig. 1.22.*

Fig. 1.22 Spacing radius

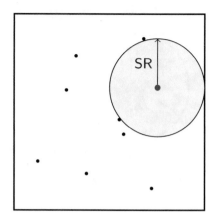

By the definition, $\mathsf{SR}(\mathbb{Z}_n) \leq \mathsf{CR}(\mathbb{Z}_n)$ for any $\mathbb{Z}_n$ as the balls used in the definition of spacing must fully belong to $\mathscr{X}$. The theory of spacings is well-developed for $d = 1$, but no non-trivial analytical results in the case $d > 1$ are known to the authors except when $z_i \in \mathbb{Z}_n$ are i.i.d. random points. Let us quote some relevant results obtained in [7, 18] and translate them to the language of spacing radius.

The maximum (multivariate) spacing of a set of points $\mathbb{Z}_n = \{z_1, \ldots, z_n\}$ with respect to a convex set $B \subset \mathbb{R}^d$ is the largest possible subset $x + tB$ of $\mathscr{X}$ which does not contain any of the points $z_i \in \mathbb{Z}_n$. Let $\mathrm{vol}(\mathscr{X}) = 1$, B be either a cube or a Euclidean ball in $\mathbb{R}^d$ of unit volume: $\mathrm{vol}(B) = 1$. Let also $z_1, \ldots, z_n$ be i.i.d. random points with the uniform distribution on $\mathscr{X}$. Set

$$\triangle_n = \sup\{t : \text{there exists } x \in \mathbb{R}^d \text{ such that } x + tB \subset (\mathscr{X} \setminus \mathbb{Z}_n)\}$$

and define the volume of the maximum spacing as $V_n = (\triangle_n)^d$, which is the volume of the largest ball (or cube of fixed orientation) that is contained in $\mathscr{X}$ and avoids all n points $z_1, \ldots, z_n$. Then we have

$$\lim_{n \to \infty} \frac{n V_n - \ln n}{\ln \ln n} = d - 1 \quad \text{with probability 1.} \tag{1.2.16}$$

Moreover, as $n \to \infty$ the sequence of random variables $n V_n - \ln n - (d-1) \ln \ln n + \beta_d$ converges in distribution to the random variable with c.d.f. $\exp(-e^{-u})$, $u > 0$, where $\beta_d = 0$ if B is a cube and

$$\beta_d = \ln \Gamma(d+1) - (d-1)\left[\frac{1}{2}\ln \pi + \ln \Gamma\left(\frac{d}{2}+1\right) - \ln \Gamma\left(\frac{d+1}{2}\right)\right] \tag{1.2.17}$$

if B is a ball. For large d, the quantity β_d defined in (1.2.17) can be approximated as

$$\beta_d = \frac{d}{2}\ln\frac{2d}{\pi} - d + \ln(\pi d) - \frac{1}{4} + O\left(\frac{1}{d}\right).$$

This approximation is very accurate already for $d \geq 5$.

From (1.2.16), for large n we have

$$\triangle_n = V_n^{1/d} \simeq \left[\frac{\ln n - (d-1)\ln\ln n}{n} \right]^{1/d}. \qquad (1.2.18)$$

Recall the approximate formula (1.1.4) for the radius of the ball of volume 1, from which $\mathsf{SR}(\mathbb{Z}_n) \simeq \triangle_n \sqrt{d}/\sqrt{2\pi e}$. Combining this with (1.2.18) we obtain for large n and reasonable large d:

$$\mathsf{SR}(\mathbb{Z}_n) \simeq \frac{\sqrt{d}}{\sqrt{2\pi e}} \left[\frac{\ln n - (d-1)\ln\ln n}{n} \right]^{1/d}. \qquad (1.2.19)$$

For large d, the values of $\mathsf{SR}(\mathbb{Z}_n)$ are clearly rather large, even if n is very large. This is caused by very slow decrease of the rhs of (1.2.19), as n grows.

1.3 Covering

This section is fully devoted to covering, which we consider as the main characteristic of space-filling.

1.3.1 Covering Radius

The covering radius CR of $\mathbb{Z}_n = \{z_1, \ldots, z_n\}$ is defined by (1.2.12). Point sets with small CR, like the one shown in Fig. 1.23, are very desirable in practice. In computer experiments, the covering radius is called minimax-distance criterion;

Fig. 1.23 A set of points with small CR

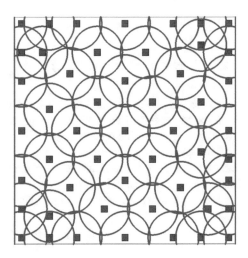

see [19] and [33]; in the theory of low-discrepancy sequences, covering radius is called dispersion; see [26, Ch. 6]. The problem of optimal covering of a cube by n balls has very high importance for the theory of global optimization and many branches of numerical mathematics. In particular, the celebrated results of A.G.Sukharev imply that an n-point design $\mathbb{Z}_n$ with smallest CR provides the following: (a) min-max n-point global optimization method in the set of all adaptive n-point optimization strategies (see [37] and [38, Ch.4,Th.2.1]), (b) worst-case n-point multiobjective global optimization method in the set of all adaptive n-point algorithms (see [43]) and (c) the n-point min-max optimal quadrature (see [38, Ch.3,Th.1.1]). In all three cases, the class of (objective) functions is the class of Lipschitz functions, and the optimality of the design is independent of the value of the Lipschitz constant. Sukharev's results on n-point min-max optimal quadrature formulas have been generalized in [30] for functional classes different from the class of Lipschitz functions; see also formula (2.3) in [10].

1.3.2   Star-Discrepancy and Covering Radius

Unlike discrepancies, which are global characteristics of uniformity measuring the distance between two probability measures (μ_n and μ; see Sect. 1.2.1), CR is a local characteristic of space-filling. For $\mathscr{X} = [0, 1]^d$, if $\mathbb{Z}_n$ has small star-discrepancy, then it cannot have very large CR. This follows from the following results, [26], pp. 150–156. Denote the L_∞-norm by $\|u - v\|_\infty = \max_{i=1,\ldots,d} |u_i - v_i|$, where $u = (u_1, \ldots, u_d), v = (v_1, \ldots, v_d) \in \mathbb{R}^d$. Then for $\mathscr{X} = [0, 1]^d$ we have (see, e.g. [26, p. 15 and 152])

$$\mathsf{CR}_\infty(\mathbb{Z}_n) \leq \mathsf{CR}(\mathbb{Z}_n) \leq \sqrt{d} \cdot \mathsf{CR}_\infty(\mathbb{Z}_n); \quad \mathsf{CR}_\infty(\mathbb{Z}_n) \leq 2\left[D_n^*\right]^{1/d} ; \quad (1.3.1)$$

and $\frac{1}{2} \leq \inf_{\mathbb{Z}_\infty} \limsup_n n^{1/d} \, \mathsf{CR}_\infty(\mathbb{Z}_n) \leq 1/\log 4$. This implies

$$(\text{any } \mathbb{Z}_n)\frac{1}{2}n^{-1/d} \leq \mathsf{CR}(\mathbb{Z}_n) \leq \frac{\sqrt{d}}{\log 4}n^{-1/d}\,(\text{optimal } \mathbb{Z}_n, n \to \infty). \quad (1.3.2)$$

In view of the results of Sect. 1.3.4, the upper bound in (1.3.2) is much sharper than the lower bound in (1.3.2) as we should expect a rather large covering radius for large d even for the optimal coverings.

The reverse is not correct: small CR does not imply small discrepancy. For example, in $d = 1$, the sequence with the smallest CR is not uniform as it has a discrepancy not tending to zero.

We can define a low-dispersion sequence $\mathbb{Z}_\infty$ as a sequence with

$$C(\mathbb{Z}_\infty) = \limsup_n n^{1/d} \, \mathsf{CR}(\mathbb{Z}_n) < \infty. \quad (1.3.3)$$

The second equality in (1.3.1) implies that any low-discrepancy sequence $\mathbb{Z}_\infty$ is low-dispersion in the sense of definition (1.3.3). There is, however, huge uncertainty related to the value of the constant $C(\mathbb{Z}_\infty)$ in (1.3.3). The value $C_* = \inf_{\mathbb{Z}_\infty} C(\mathbb{Z}_\infty)$ is known only in the case $d = 1$; it is $C_* = 1/\log 4$; see [26, p. 154]. For $d > 1$, in view of (1.3.2), $C_* \leq \sqrt{d}/\log 4$ and, as discussed above, the true value of C_* should be rather close to $\sqrt{d}/\log 4$. We can thus modify the definition above of the low-dispersion sequence by requiring

$$C(\mathbb{Z}_\infty) \leq \sqrt{d}/\log 4 \qquad (1.3.4)$$

rather than $C(\mathbb{Z}_\infty) < \infty$ in (1.3.3). In this case, as far as the authors aware, all known low-discrepancy sequences are no longer low-dispersion. Proof of existence of a low-dispersion sequence in the sense of definition (1.3.4) is given in [26], p. 155. We believe that there is no sequence such that the strict inequality in (1.3.4) can be achieved, for any d.

Conjecture *Let $\mathcal{X}$ be any measurable subset of $\mathbb{R}^d$ with $0 < \mathrm{vol}(\mathcal{X}) < \infty$. There is no sequence $\mathbb{Z}_\infty = \{z_1, z_2, \ldots\} \subset \mathcal{X}$ with $C(\mathbb{Z}_\infty) < \sqrt{d}/\log 4$.*

1.3.3 Covering Radius and Weighted Discrepancy

The connection between star-discrepancy and covering radius is even stronger when considering design measures and weighted discrepancies. Consider the case $d = 1$, and let ξ_n be a probability measure supported on $\mathbb{Z}_n$ with weight w_i on z_i; assume, without any loss of generality, that $0 \leq z_1 < z_2 < \cdots < z_n \leq 1$. The weighted star-discrepancy of ξ_n is defined as $D^*(\xi_n) = \sup_{0 \leq t < 1} \left| \sum_{i:\, z_i \leq t} w_i - t \right|$. The covering radius of $\mathbb{Z}_n$ is $\mathrm{CR}(\mathbb{Z}_n) = \max\{z_1, (z_2 - z_1)/2, \ldots, (z_n - z_{n-1})/2, 1 - z_n\}$. For simplicity, we restrict our attention to designs with $z_1 = 0$ and $z_n = 1$. We then have the following result.

Theorem 1.2 *For any design ξ_n such that $0 = z_1 < z_2 < \cdots < z_{n-1} < z_n = 1$, we have (i) $\mathrm{CR}(\mathbb{Z}_n) = D^*(\xi_n^*)$, where ξ_n^* has the weights $w_1^* = z_2/2$, $w_n^* = (1 - z_{n-1})/2$ and $w_i^* = (z_{i+1} - z_{i-1})/2$ for $i = 2, \ldots, n-1$ and (ii) $D^*(\xi_n) > \mathrm{CR}(\mathbb{Z}_n)$ for any other probability measure ξ_n supported on $\mathbb{Z}_n$.*

Proof One can check that, for any design ξ_n supported on $\mathbb{Z}_n$,

$$D^*(\xi_n) = \max_{1 \leq i \leq n} \left\{ \frac{w_i}{2} + \left| z_i - \frac{W_i + W_{i-1}}{2} \right| \right\}, \qquad (1.3.5)$$

where $W_0 = 0$ and $W_i = w_1 + \cdots + w_i$ for $i = 1, \ldots, n$. This expression is a generalization of that in [26, Theorem 2.6] for the classical star-discrepancy. It is then straightforward to check that $D^*(\xi_n^*) = \mathsf{CR}(\mathbb{Z}_n)$. Moreover, if we take $\xi_n = \xi_n^*$, then all the terms in the right-hand side of (1.3.5) are equal to $\mathsf{CR}(\mathbb{Z}_n)$:

$$\frac{1}{2} w_i^* + \left| z_i - \frac{1}{2}(W_i^* + W_{i-1}^*) \right| = \mathsf{CR}(\mathbb{Z}_n), \quad i = 1, \dots, n.$$

This implies that for any other set of weights $w_1, \dots, w_n$; we have $D^*(\xi_n) > \mathsf{CR}(\mathbb{Z}_n)$. ∎

1.3.4 Weak Covering

The main problem discussed in this section is the following problem of covering a cube by n balls. Let $[-1, 1]^d$ be a d-dimensional cube, $z_1, \dots, z_n \in \mathbb{R}^d$ and $\mathscr{B}_d(z_j, r)$ be the corresponding balls of radius r centred at z_j ($j = 1, \dots, n$). The dimension d, the number of balls n and their radius r could be arbitrary.

We are interested in the problem of choosing the locations of the centres of the balls $z_1, \dots, z_n$ so that the union of the balls $\cup_j \mathscr{B}_d(z_j, r)$ covers the largest possible proportion of the cube $[-1, 1]^d$. That is, we are interested in choosing a scheme (a collection of points) $\mathbb{Z}_n = \{z_1, \dots, z_n\}$ so that

$$C_d(\mathbb{Z}_n, r) := \mathrm{vol}([-1, 1]^d \cap \mathscr{B}_d(\mathbb{Z}_n, r))/2^d \tag{1.3.6}$$

is as large as possible (given n, r and the freedom we are able to use in choosing $z_1, \dots, z_n$). Here

$$\mathscr{B}_d(\mathbb{Z}_n, r) = \bigcup_{j=1}^{n} \mathscr{B}_d(z_j, r) \tag{1.3.7}$$

and $C_d(\mathbb{Z}_n, r)$ is the proportion of the cube $[-1, 1]^d$ covered by the balls $\mathscr{B}_d(z_j, r)$ ($j = 1, \dots, n$).

If d is not small (say, $d > 5$), then computation of the covering radius $\mathsf{CR}(\mathbb{Z}_n)$ for any non-trivial design $\mathbb{Z}_n$ is a very difficult computational problem. This explains why the problem of construction of optimal n-point designs with smallest covering radius is notoriously difficult; see, for example, recent surveys [41, 42].

If $r = \mathsf{CR}(\mathbb{Z}_n)$, then $C_d(\mathbb{Z}_n, r)$ defined in (1.3.6) is equal to 1, and the whole cube $[-1, 1]^d$ gets covered by the balls. However, we are only interested in reaching values like 0.95, when a large part of the ball is covered. There are two main reasons why we are not interested in reaching the value $C_d(\mathbb{Z}_n, r) = 1$: (a) practical impossibility of making a numerical check of the full coverage, if d is large enough, and (b) our approximations lose accuracy when $C_d(\mathbb{Z}_n, r)$ closely approaches 1.

If, for a given $\gamma \in [0, 1)$, we have $C_d(\mathbb{Z}_n, r) \geq 1 - \gamma$, then the corresponding coverage of $[-1, 1]^d$ will be called $(1 - \gamma)$-coverage; the corresponding value of r

can be called $(1-\gamma)$-covering radius. If $\gamma = 0$ then the $(1-\gamma)$-coverage becomes the full coverage, and 1-covering radius of $\mathbb{Z}_n$ becomes $C_d(\mathbb{Z}_n, r)$. Of course, for any $\mathbb{Z}_n = \{z_1, \ldots, z_n\}$ we can reach $C_d(\mathbb{Z}_n, r) = 1$ by means of increasing r. Likewise, for any given r, we can reach $V_d(\mathbb{Z}_n, r) = 1$ by sending $n \to \infty$. However, we are not interested in very large values of n and try to get the coverage of the most part of the cube $[-1, 1]^d$ with the radius r as small as possible. We will keep in mind the following typical values of d and n: $d = 10, 20, 50$ and $n = 64, 128, 512, 1024$. Correspondingly, we will illustrate our results for such scenarios.

The main covering scheme The following will be our main scheme for choosing $\mathbb{Z}_n = \{z_1, \ldots, z_n\}$: $z_1, \ldots, z_n$ are i.i.d. random vectors uniformly distributed in the cube $[-\delta, \delta]^d$, where $\delta \in [0, 1]$ is a parameter.

The reasons why we have chosen this scheme are:

– it is easier to theoretically investigate than all other non-trivial schemes;
– it includes, as a special case $\delta = 1$, the scheme which is very popular in practice of Monte Carlo [26] and global random search (PRS; see Chap. 3);
– numerical studies show that the coverings with suitable values of δ are rather efficient, especially for large d.

Theoretical investigation Let $z_1, \ldots, z_n$ be i.i.d. random vectors uniformly distributed in the cube $[-\delta, \delta]^d$ with $0 < \delta \le 1$. Then, for given $U = (u_1, \ldots, u_d)^\top \in \mathbb{R}^d$,

$$\mathbb{P}\{U \in \mathscr{B}_d(\mathbb{Z}_n, r)\} = 1 - \prod_{j=1}^{n} \mathbb{P}\{U \notin \mathscr{B}_d(z_j, r)\}$$

$$= 1 - \prod_{j=1}^{n}\left(1 - \mathbb{P}\{U \in \mathscr{B}_d(z_j, r)\}\right) = 1 - \left(1 - \mathbb{P}_z\{\|U - Z\| \le r\}\right)^n, \quad (1.3.8)$$

where $\mathscr{B}_d(\mathbb{Z}_n, r)$ is defined in (1.3.7). The main characteristic of interest $C_d(\mathbb{Z}_n, r)$, defined in (1.3.6), the proportion of the cube covered by the union of balls $\mathscr{B}_d(\mathbb{Z}_n, r)$, is simply

$$C_d(\mathbb{Z}_n, r) = \mathbb{E}_U \mathbb{P}\{U \in \mathscr{B}_d(\mathbb{Z}_n, r)\}. \quad (1.3.9)$$

We shall simplify the expression (1.3.8) with the help of the approximation $(1 - t)^n \simeq e^{-nt}$, which is very accurate for small values of t and moderate values of nt; this agrees with the ranges of d, n and r we are interested in. Continuing (1.3.8), note that

$$\mathbb{P}_z\{\|U - Z\| \le r\} = \mathbb{P}_z\left\{\sum_{j=1}^{d}(z_j - u_j)^2 \le r^2\right\} = C_{d,U,r}^{(\delta)},$$

where $C_{d,U,r}^{(\delta)}$ is defined by the formula (1.1.9). From (1.1.10) we have $C_{d,U,r}^{(\delta)} = C_{d,U/\delta,r/\delta}$ where $C_{d,U/\delta,r/\delta}$ is the quantity defined by (1.1.8). This quantity can be approximated in a number of different ways as shown in Sect. 1.1.8. We will compare (1.1.13), the simplest of the approximations, with the approximation given in (1.1.16). Approximation (1.1.13) gives

$$C_{d,U,r}^{(\delta)} = C_{d,U/\delta,r/\delta} \cong \Phi\left(\frac{(r/\delta)^2 - \|U\|^2/\delta^2 - d/3}{2\sqrt{\|U\|^2/(3\delta^2) + d/45}}\right), \qquad (1.3.10)$$

whereas approximation (1.1.16) yields

$$C_{d,U,r}^{(\delta)} \cong \Phi(t_\delta) + c_d \frac{\|U\|^2/\delta^2 + d/63}{5\sqrt{3}(\|U\|^2/\delta^2 + d/15)^{3/2}}(1 - t_\delta^2)\phi(t_\delta), \qquad (1.3.11)$$

with $c_d = 1 + 4/d$ and

$$t_\delta = \frac{(r/\delta)^2 - \|U\|^2/\delta^2 - d/3}{2\sqrt{\|U\|^2/(3\delta^2) + d/45}}.$$

It is easy to compute $\mathbb{E}\|U\|^2 = d/3$ and $\mathrm{var}(\|U\|^2) = 4d/45$. Moreover, if d is large enough, then $\|U\|^2 = \sum_{j=1}^{d} u_j^2$ is approximately normal.

We can combine the expressions (1.3.9) and (1.3.8) with approximations (1.3.10), (1.3.11) and $(1 - t)^n \simeq e^{-nt}$, as well as with the normal approximation for the distribution of $\|U\|^2$, to arrive at two final approximations for $C_d(\mathbb{Z}_n, r)$ that differ in complexity. If the original normal approximation of (1.3.10) is used, then we obtain

$$C_d(\mathbb{Z}_n, r) \simeq 1 - \int_{-\infty}^{\infty} \psi_1(s)\phi(s)ds, \qquad (1.3.12)$$

with

$$\psi_1(s) = \exp\{-n\Phi(c_s)\}, \quad c_s = \frac{3(r/\delta)^2 - s' - d}{2\sqrt{s' + d/5}}, \quad s' = (d + 2s\sqrt{d/5})/\delta^2.$$

If approximation (1.3.11) is used, we obtain:

$$C_d(\mathbb{Z}_n, r) \simeq 1 - \int_{-\infty}^{\infty} \psi_2(s)\phi(s)ds, \qquad (1.3.13)$$

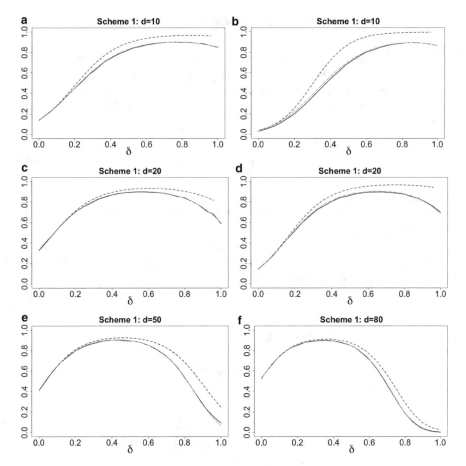

Fig. 1.24 (a) $C_d(\mathbb{Z}_n, r)$ and approximations: $n = 128$. (b) $C_d(\mathbb{Z}_n, r)$ and approximations: $n = 512$. (c) $C_d(\mathbb{Z}_n, r)$ and approximations: $n = 128$. (d) $C_d(\mathbb{Z}_n, r)$ and approximations: $n = 512$. (e) $C_d(\mathbb{Z}_n, r)$ and approximations: $n = 512$. (f) $C_d(\mathbb{Z}_n, r)$ and approximations: $n = 512$

with

$$\psi_2(s) = \exp\left\{-n\left(\Phi(c_s) + \left(1 + \frac{4}{d}\right)\frac{s' + d/21}{5[s' + d/5]^{3/2}}(1 - c_s^2)\phi(c_s)\right)\right\}.$$

Simulation study for assessing accuracy of approximations (1.3.12) and (1.3.13) In Fig. 1.24, $C_d(\mathbb{Z}_n, r)$ is represented by a solid black line and has been obtained via Monte Carlo methods. Approximation (1.3.12) is indicated by a dashed blue line, and approximation (1.3.13) is represented by long dashed green lines. All figures demonstrate that approximation (1.3.13) is extremely accurate across different dimensions and values of n. This approximation is much superior to approximation (1.3.12).

1.3.5 Relation Between Quantization and Weak Covering

Let us return to the mean squared quantization error $QE(\mathbb{Z}_n)$ defined by (1.2.13).

The two characteristics, $C_d(\mathbb{Z}_n, r)$ and $QE(\mathbb{Z}_n)$, are related as follows: $C_d(\mathbb{Z}_n, r)$, as a function of $r \geq 0$, is the c.d.f. of the r.v. $\varrho(X, \mathbb{Z}_n) = \min_{z_i \in \mathbb{Z}_n} \|X - z_i\|$, where X is uniform on $\mathscr{X}$, while $QE(\mathbb{Z}_n)$ is the second moment of the distribution with this c.d.f.:

$$QE(\mathbb{Z}_n) = \int_{r \geq 0} r^2 dC_d(\mathbb{Z}_n, r) . \tag{1.3.14}$$

This implies that any approximations for $C_d(\mathbb{Z}_n, r)$ automatically create corresponding approximations for the mean squared quantization error $QE(\mathbb{Z}_n)$. This topic is explored in [28] and [29].

Note that the designs creating the so-called centroidal Voronoi tessellations, where the design points z_i become the centres of masses of the Voronoi cells, possess extra optimality properties; see [10, 11].

1.3.6 Regularized Covering

By approximating L_∞-norms with L_q-norm ($q \neq 0$) in (1.2.12), we get the so-called regularized covering radius:

$$CR^{(q)}(\mathbb{Z}_n) = \left[\int \left(\frac{1}{n} \sum_{i=1}^n \|z_i - x\|^{-q} \right)^{-1} \mu(dx) \right]^{1/q} ,$$

where μ is the uniform measure on $\mathscr{X}$. For any $\mathbb{Z}_n$, $\Phi(\mathbb{Z}_n) = \lim_{q \to \infty} CR^{(q)}(\mathbb{Z}_n)$.

Continuous version of the criterion $CR^{(q)}(\mathbb{Z}_n)$

$$\phi_q(\xi) = \left[\int \frac{1}{\int \|x - z\|^{-q} \xi(dz)} \mu(dx) \right]^{1/q} \quad q \neq 0.$$

so that $CR^{(q)}(\mathbb{Z}_n) = \phi_q(\mu_n)$.

As shown in [34], $\phi_q^q(\xi)$ is a convex functional and hence can be considered as an optimality criterion for construction of optimal designs; n-point approximations to these optimal designs provide the schemes similar to those that minimize the regularized covering $CR^{(q)}(\mathbb{Z}_n)$; see [34] for details and algorithms.

1.4 Bibliographic Notes

Section 1.1 More material for topics touched upon in Sects. 1.1.1, 1.1.2, 1.1.3, 1.1.4, 1.1.5, and 1.1.6 can be found in [1, 2]. Section 1.1.8 follows [27]. Results presented in Sect. 1.1.8 have been generalized in [29] for a more general class of sampling distributions.

Section 1.2 Much more material on standard discrepancies and low-discrepancy sequences can be found in many references; see, e.g. [9, 16, 25, 26]. We refer to [35] for a comprehensive survey on space-filling and kernel discrepancies. The concept of weighted discrepancy discussed in Sect. 1.2.2 is different from the one used in [24] and similar publications, where the dimensions, rather than points, may have different weights. We use the definition (1.15) with $r = 1$ from [39] where we assume that the points can be assigned some weights. Section 1.2.3: I have been discussing the idea of looking (proving) at the Riemann hypothesis via the discrepancies of the weighted Farey sequences with many eminent number theorists (most notably with Martin Huxley), but this idea is clearly formulated for the first time here. Material of Sect. 1.2.4 is new. It is based on an unpublished manuscript of Henry P. Wynn (London School of Economics and Alan Turing Institute) and A. Zhigljavsky. Some additional references for the material of Sect. 1.2.5 are as follows: for quantization in space, see [4–6] and Chapter 33 in [14]; for relation between quantization in a cube and covering, see [28]; for packing in space, see [6, 41, 42]; and for construction of Fekete points, see, e.g. [3, 15, 36].

Section 1.3 For covering of space, see [6, 41]. For general properties of the covering radius (called dispersion) for low-discrepancy sequences, see [26, Chapter 5]. Material of Sect. 1.3.3 is new; see also [35]. Section 1.3.4 follows [27]. Results presented in Sect. 1.3.4 have been significantly generalized and improved in [28, 29]. Approximations for the mean squared quantization error mentioned in Sect. 1.3.5 are constructed in [28, 29]. As the principal reading on quantization, we recommend [13]. The concept of regularized covering introduced in Sect. 1.3.6 is studied in [34].

References

1. K. Ball, An elementary introduction to modern convex geometry. Fla. Geogr. **31**, 1–58 (1997)
2. A. Blum, J. Hopcroft, R. Kannan, *Foundations of Data Science* (Cambridge University Press, 2020)
3. L. Bos, S. De Marchi, A. Sommariva, M. Vianello, Computing multivariate Fekete and Leja points by numerical linear algebra. SIAM J. Numer. Anal. **48**(5), 1984–1999 (2010)
4. J. Conway, N. Sloane, Fast quantizing and decoding and algorithms for lattice quantizers and codes. IEEE Trans. Inf. Theory **28**(2), 227–232 (1982)
5. J. Conway, N. Sloane, Voronoi regions of lattices, second moments of polytopes, and quantization. IEEE Trans. Inf. Theory **28**(2), 211–226 (1982)
6. J.H. Conway, N.J.A. Sloane, *Sphere Packings, Lattices and Groups*, 3rd edn. (Springer, 1999)

7. P. Deheuvels, Strong bounds for multidimensional spacings. Z. Wahrsch. Verw. Gebiete **64**, 411–424 (1983)
8. F. Dress, Discrépance des suites de Farey. J. de théorie des nombres de Bordeaux. **11**(2), 345–367 (1999)
9. M. Drmota, R.F. Tichy, *Sequences, Discrepancies and Applications* (Springer, Berlin, 1997)
10. Q. Du, V. Faber, M. Gunzburger, Centroidal voronoi tessellations: Applications and algorithms. SIAM Rev. **41**(4), 637–676 (1999)
11. Q. Du, M. Gunzburger, L. Ju, Advances in studies and applications of centroidal voronoi tessellations. Numer. Math. Theory Meth. Appl. **3**(2), 119–142 (2010)
12. J. Franel, Les suites de Farey et le problème des nombres premiers. *Nachrichten von der Gesellschaft der Wissenschaften zu Göttingen, Mathematisch-Physikalische Klasse* **1924**, 198–201 (1924)
13. S. Graf, H. Luschgy, *Foundations of Quantization for Probability Distributions* (Springer, 2007)
14. P. Gruber, *Convex and Discrete Geometry* (Springer, 2007)
15. D.P. Hardin, E.B. Saff, Discretizing manifolds via minimum energy points. Not. AMS **51**(10), 1186–1194 (2004)
16. F.J. Hickernell, A generalized discrepancy and quadrature error bound. Math. Comput. **67**(221), 299–322 (1998)
17. M. Huxley, A. Zhigljavsky, On the distribution of Farey fractions and hyperbolic lattice points. Period. Math. Hung. **42**(1–2), 191–198 (2001)
18. S. Janson, Maximal spacings in several dimensions. Ann. Probab. **15**, 274–280 (1987)
19. M.E. Johnson, L.M. Moore, D. Ylvisaker, Minimax and maximin distance designs. J. Statist. Plan. Inference **26**(2), 131–148 (1990)
20. S. Kanemitsu, M. Yoshimoto, Farey series and the Riemann hypothesis. Acta Arithmetica **75**(4), 351–374 (1996)
21. S. Kanemitsu, M. Yoshimoto, Farey series and the Riemann hypothesis III. Ramanujan J. **1**(4), 363–378 (1997)
22. P. Kargaev, A. Zhigljavsky, Asymptotic distribution of the distance function to the Farey points. J. Number Theory **65**(1), 130–149 (1997)
23. E. Landau, *Bemerkungen zu der vorstehenden Abhandlung von Herrn Franel* (1924)
24. G. Larcher, F. Pillichshammer, K. Scheicher, Weighted discrepancy and high-dimensional numerical integration. BIT Numer. Math. **43**(1), 123–137 (2003)
25. J. Matoušek, *Geometric Discrepancy: An Illustrated Guide* (Springer, 1999)
26. H. Niederreiter, *Random Number Generation and Quasi-Monte Carlo methods* (SIAM, Philadelphia, 1992)
27. J. Noonan, A. Zhigljavsky, Covering of high-dimensional cubes and quantization. SN Operations Research Forum **1**(3), 1–32 (2020)
28. J. Noonan, A. Zhigljavsky, Efficient quantization and weak covering of high dimensional cubes. arXiv:2005.07938 (2020)
29. J. Noonan, A. Zhigljavsky, Non-lattice covering and quanitization of high dimensional sets, in *Black Box Optimization, Machine Learning and No-Free Lunch Theorems*, volume arXiv:2006.02705 (Springer, 2021)
30. G. Pagès, A space quantization method for numerical integration. J. Comput. Appl. Math. **89**(1), 1–38 (1998)
31. V.V. Petrov, *Sums of Independent Random Variables* (Springer, 1975)
32. V.V. Petrov, *Limit Theorems of Probability Theory: Sequences of Independent Random Variables* (Oxford Science Publications, 1995)
33. L. Pronzato, W.G. Müller, Design of computer experiments: space filling and beyond. Stat. Comput. **22**(3), 681–701 (2012)
34. L. Pronzato, A. Zhigljavsky, Measures minimizing regularized dispersion. J. Sci. Comput. **78**(3), 1550–1570 (2019)
35. L. Pronzato, A. Zhigljavsky, Bayesian quadrature, energy minimization, and space-filling design. SIAM/ASA J. Uncertain. Quantification **8**(3), 959–1011 (2020)

36. E.B. Saff, A.B. Kuijlaars, Distributing many points on a sphere. Math. Intelligencer **19**(1), 5–11 (1997)
37. A.G. Sukharev, Optimal strategies of search for an extremum. USSR Comput. Math. Math. Phys. **11**(4), 910–924 (1971)
38. A.G. Sukharev, *Minimax Models in the Theory of Numerical Methods* (Springer, 1992)
39. V. Temlyakov, Cubature formulas, discrepancy, and nonlinear approximation. J. Complexity **19**(3), 352–391 (2003)
40. B. Tibken, D. Constales, et al., The volume of the intersection of a concentric cube and ball in n-dimensional space: collection of approximations. SIAM Rev. **39**, 783–786 (1997)
41. G.F. Tóth, Packing and covering, in *Handbook of Discrete and Computational Geometry* (Chapman and Hall/CRC, 2017), pp 27–66
42. G.F. Tóth, W. Kuperberg, Packing and covering with convex sets, in *Handbook of Convex Geometry* (Elsevier, 1993), pp 799–860
43. A. Žilinskas, On the worst-case optimal multi-objective global optimization. Optim. Lett. **7**(8), 1921–1928 (2013)

Chapter 2
Bi-objective Decisions and Partition-Based Methods in Bayesian Global Optimization

We present in this chapter our recent work in Bayesian approach to continuous non-convex optimization. A brief review precedes the main results to have our work presented in the context of challenges of the approach.

Non-convex continuous optimization problems are frequent in various applications. Black-box problems is difficult to solve, but important subclass of those problems where the objective function is not defined by a formula but software is available to compute its values. The analytical properties of black-box functions, besides continuity, usually cannot be substantiated. The possibility of the unfavourable properties of black-box functions such as non-differentiability and multimodality should not be ignored. The solution of black-box problems is further complicated when the objective function is computational intensive. Such problems are commonly referred to as expensive black-box problems. The uncertainty in properties of the objective functions and their expensiveness complicate the development of the appropriate optimization algorithms. We address that challenge based on the theory of rational decision-making under uncertainty. Specifically, we consider the optimization process as a sequence of decision-making under uncertainty. The selection of a model of uncertainty is a counterpart of the definition of the rational decision. Stochastic functions are natural models of black-box objective functions. We are interested in the algorithm which is average optimal with respect to the properly selected stochastic function. A classical example of the optimality criterion is the average error of the approximation of the global minimum.

The algorithm with minimum average error of the approximation of global minimum was proposed and named *Bayesian method for seeking an extremum* in [78]. It became common later to apply the attribute Bayesian for a global optimization (GO) algorithm which is in some sense optimal with respect to a probabilistic model of the objective functions. The concept of average case optimality is important for the general theory of algorithms as well as for the development of efficient methods for the solution of various mathematical and information processing problems [27, 86, 102].

© The Author(s) 2021
A. Zhigljavsky, A. Žilinskas, *Bayesian and High-Dimensional Global Optimization*,
SpringerBriefs in Optimization, https://doi.org/10.1007/978-3-030-64712-4_2

To our best knowledge, the first Bayesian GO algorithm was proposed by Kushner [63] much earlier than this term was introduced. The proposed algorithm is aimed at optimization of a univariate multiextremal function, given that the only information available is the noise-perturbed samples of the function. The point for the current computation of the objective function is defined by maximizing the probability to improve the best value found so far. The improvement probability is computed using the Wiener process as a probabilistic model of the objective function. The idea of this algorithm is described as *an attempt to avoid the optimal search problem through the use of several intuitively reasonable heuristics*. The common idea of [63] and [78] is in the modelling of the objective function by a stochastic function. An optimal algorithm is considered in both cases although optimality criteria are different.

The first theoretical results on Bayesian GO as well as the experimental testing results were encouraging. However, only few papers were published on this subject during two decades after the original publications, possibly because of the following reasons. Two main problems should be solved to implement the theoretical advantages of the average case optimality. First, an appropriate probability distribution on the set of potential problems should be defined. Second, a computer code of acceptable complexity should be developed that implements the theoretical method with sufficient accuracy. These problems remain a challenge from the very beginning of the research in Bayesian GO to this day. We discuss in this chapter our attempts to cope with the mentioned challenges. A brief review of the related publications is included to present our results in the context of the contemporary research in that area. It is important to note that we focus on the algorithms for the optimization of continuous non-convex functions, while the term Bayesian GO is used in a broader sense, e.g. in the Bandit theory and corresponding applications.

2.1   Original Algorithms

We start the presentation of the original algorithms with the optimal Bayesian algorithm proposed in [78] because of two reasons. First, the name of this algorithm has become generic. Second, the earlier proposed algorithm [63] is designed to solve special problems, i.e. for the minimization of univariate functions in the presence of noise.

A black-box minimization problem $\min_{x \in \mathbf{A}} f(x)$, $\mathbf{A} \subset R^d$, is considered, where $f(\cdot)$ is a continuous function and $\mathbf{A}$ is a compact set with a nonempty interior. Let's assume that the uncertainty about $f(x)$ can be modelled by a stochastic function. Recall that stochastic functions of single variable are normally called *stochastic/random processes* and the term *random field* is used in the general case. Let a random field $\xi(x)$, $x \in \mathbf{A}$, be accepted as a probabilistic model of the objective function. The selection of an appropriate model can be quite complicated as will be discussed in Sect. 2.4.

Let us assume that the number of function samples N is fixed in advance. An algorithm can be described by the vector-function $\mathscr{A} = (\mathscr{A}_1(\cdot), \ldots, \mathscr{A}_{N+1}(\cdot))^T$, where $\mathscr{A}_i(\cdot)$, $i = 2, \ldots, N$, defines a point x_i for computing the objective function value at iteration i:

$$x_i = \mathscr{A}_i(x_j, y_j, j = 1, \ldots, i - 1), i = 2, \ldots, N, \tag{2.1.1}$$

$\mathscr{A}_1(\cdot)$ defines the point of the first computation x_1 depending on the used probabilistic model and on N; $\mathscr{A}_{N+1}(\cdot)$ defines the approximation of the global minimizer $\tilde{x}_{oN}$ depending on all available information at the end of search. The best found point frequently is accepted as $\tilde{x}_{oN}$.

We have assumed that the algorithm $\mathscr{A}$ is applied to minimize the random field $\xi(x)$. That assumption implies that the values $y_i = \xi(x_i)$ in (2.1.1), as well as approximation of the global minimum $\xi(\tilde{x}_{oN})$, are random variables. Thus, the average error of the approximation of the global minimum is a natural criterion of the efficiency of $\mathscr{A}$. For general assumptions on $\xi(x)$ and $\mathscr{A}_i(\cdot)$, there exists the algorithm $\tilde{\mathscr{A}}$ with the minimum average error

$$\tilde{\mathscr{A}} = \arg \min_{\mathscr{A}} \mathbb{E}(\xi(\tilde{x}_{oN}) - \min_{x \in \mathbf{A}} \xi(x)), \tag{2.1.2}$$

which is named Bayesian in [78]. Since the term *Bayesian algorithm* was later used in a broader sense, we will refer to the algorithm $\tilde{\mathscr{A}}$ as *the optimal Bayesian algorithm*. Dynamic programming is applied to prove the existence of $\tilde{\mathscr{A}}$ in [78]. Theoretically, the optimal Bayesian algorithm can also be implemented using the dynamic programming approach. However, the high computational complexity of that algorithm hinders its practical implementation. Thus, weakening of the requirement of optimality seemed reasonable. As a simplified version of the optimal Bayesian algorithm, a one-step Bayesian algorithm was proposed. A current iteration of the one-step Bayesian algorithm is defined as the last iteration of $\tilde{\mathscr{A}}$. The average optimal approximation of the minimum is

$$\tilde{y}_{oN} = v(x_1, \ldots, x_N, y_1, \ldots, y_N) =$$
$$= \min_{x \in \mathbf{A}} \mathbb{E}\{\xi(x) | x_j, y_j, j = 1, \ldots, N\}. \tag{2.1.3}$$

For some stochastic models, e.g. for the Wiener process, the equality $\tilde{y}_{on} = y_{on}$, $n = 1, \ldots, N$, is valid, where

$$y_{on} = \min\{y_i, i = 1, \ldots, n\}.$$

The last point x_N is chosen by $\tilde{\mathscr{A}}$ minimizing the conditional (with respect to available data) expectation of the approximation of global minimum $\tilde{y}_{oN}$

$$x_N = \arg \min_{x \in \mathbf{A}} \mathbb{E}\{v(x_1, \ldots, x_{N-1}, x, y_1, \ldots, y_{N-1}, \xi(x)) \mid$$
$$x_j, y_j, j = 1, \ldots, N - 1\}. \tag{2.1.4}$$

The iteration n of the one-step Bayesian algorithm is defined by (2.1.4) where N is replaced by n. The implementation of that algorithm is simplified in the case when $\tilde{y}_{o,n-1} = y_{o,n-1}$:

$$x_n = \arg\max_{x \in \mathbf{A}} \mathbb{E}(\max\{y_{o,n-1} - \xi(x),\ 0\} \mid x_j, y_j, j = 1, \ldots, n-1). \quad (2.1.5)$$

One of the first Bayesian algorithms is based on the Wiener process model; the convergence of this algorithm is proved for continuous objective functions, and its advantage over the algorithms known at that time is shown by the testing in [148]. A direct generalization of this algorithm to the problems of several variables is not possible since the known generalizations of the Wiener process to multidimensional space are not appropriate as models of the objective function. A solution for the use a probabilistic model is homogeneous isotropic Gaussian random field (GRF) presented in [80], where a one-step Bayesian algorithm for functions of several variables is developed. However, the generalization of the convergence proof from [148] to the algorithm [80] is difficult. Another simplified version of the optimal Bayesian algorithm is an optimal algorithm with restricted memory [164]. References to the related papers in Russian can be found in [79, 113, 145, 155].

The best found value y_{oN} is normally accepted as the approximation of the global minimum regardless of validity of the equality $\tilde{y}_{oN} = y_{oN}$ since in applied optimization problems the deterministic outcome is required. The acceptance of this approximation simplifies the computation of x_n as follows from the comparison of (2.1.4) with (2.1.5).

Recently, the new names were introduced for the one-step Bayesian algorithm and its modifications: Maximum Expected Improvement (MEI), myopic and efficient global optimization (EGO); see, e.g. [12, 57, 61]. The name MEI is used most frequently.

As mentioned above, the first Bayesian GO algorithm was proposed in [63]. This univariate algorithm for optimization in the presence of noise is constructed using the Wiener process for a probabilistic model of objective functions. For the computation of the objective function value, the algorithm selects the maximizer of the improvement probability

$$x_n = \arg\max_{x \in \mathbf{A}} \mathbf{P}\{\xi(x) \le y_{o,n-1} - \varepsilon_n \mid x_j, y_j, j = 1, \ldots, n-1\}, \quad (2.1.6)$$

where $\varepsilon_n > 0$ is an improvement threshold. This algorithm (later named the P-algorithm) was substantiated axiomatically in [154].

The Wiener process proved appropriate for the implementation of algorithms for univariate GO. The results of MEI and of the P-algorithm were promising [148, 152]. Thanks to these results, the Wiener process attracted interest of researchers as an appropriate probabilistic model for the investigation of complexity and convergence not only of GO but also of other numerical algorithms [102]. However, the Wiener process as a model of practical objective functions has several theoretical disadvantages. The estimation of the variance parameter of the

Wiener process is a serious problem which is mentioned in [63]; the problem of estimating parameters of probabilistic models in a broader context is considered in Sect. 2.4. The sample functions of the Wiener process are not differentiable almost everywhere with probability 1; thus they are quite different from the objective functions of applied problems. This problem can be circumvented by implementing a dual model. The Wiener process, as a global counterpart of the dual model, is used for the interpretation of the global behaviour of $f(x)$. A unimodal smooth function, as a local counterpart, can be used by local optimization in the subintervals which are indicated during the global search as potentially containing a global minimizer. Such hybridization is theoretically based in [149, 150].

A serious disadvantage of the MEI algorithm is the one-step planning horizon. The one-step optimality of an algorithm does not guarantee search rationality and can imply that the strategy becomes too local; this disadvantage of the one-step Bayesian algorithm has been already highlighted in [148].

An auxiliary maximization problem should be solved at every iteration of both algorithms (2.1.5) and (2.1.6). For the P-algorithm based on the Wiener process model, the maximum in (2.1.6) is defined analytically, and the computational complexity of iteration n is $O(n)$. As shown in [148], the maximization problem (2.1.5) for the Wiener process model at iteration n is reducible to $n - 2$ unimodal maximization problems. The maximization problems in (2.1.5) and (2.1.6) are non-convex. The corresponding maximization problems are also difficult due to the complexity of the computation of the expected improvement/probability. The complexity increases with n because of the involvement of the inversion of $n \times n$ ill-conditioned covariance matrices. Therefore, some heuristics are used to cope with these, strictly speaking, non-tractable problems.

Several challenges, which have been sketched here, remain in our opinion not fully addressed despite general interest in Bayesian GO. We think that the following problems still deserve attention of researchers: (1) the questionable rationality of one-step optimal (myopic) algorithms, (2) the ambiguity in the selection of a probabilistic model of the objective functions and (3) the inner computational complexity of algorithms of that approach. The contemporary topics of research are related in one or another way to the mentioned challenges; see the brief review in the next section. Our attempts to cope with these challenges are discussed in Sects. 2.3, 2.4 and 2.5.

2.2 A Review of the Contemporary Methods

The number of publications on Bayesian GO has grown in the last years rapidly. The extent of research and applications of Bayesian GO methods is well illustrated by about 24 million links provided by Google[1] to the query *Bayesian global opti-*

---

[1] About 24,000,000 results (0.56 s), accessed September 28, 2020.

mization. We briefly review in this section the contemporary research in Bayesian GO so that our results subsequently are presented in the context of the other research topics. Bayesian GO is reviewed from the perspective of other authors, e.g. in [3, 28, 50, 111]. During the last decade, several tutorials on the subject were published; see, e.g. [10, 129] and on the Internet: https://machinelearningmastery. com/what-is-bayesian-optimization/.

Recent publications frequently use a slightly different terminology and interpretation in presenting theory and algorithms of Bayesian GO optimization. The term *Gaussian process* is used for both Gaussian stochastic functions of one and many variables. The conditional distribution of a value of the underlying stochastic function with respect to the observed values is called the posterior distribution of a random variable given values of correlated observations. The computations of the objective function values by the optimization algorithm are interpreted as a sequence of observations of correlated random variables. The observation results are used for updating the probability distribution of the subsequent observations. The unknown value $f(x)$ at iteration n is modelled by a random variable with the distribution conditioned on the results of observations at previous iterations. A criterion for the selection of x is defined as a functional of the posterior distribution of the random variable which models $f(x)$. Various functionals, which are frequently called *acquisition functions*, are defined and used for the implementation of GO algorithms. This interpretation of the optimization process highlights its correspondence to the Bayesian paradigm.

The basic assumption about the aimed optimization problems is made by the definition of a probability space related to the set of potential objective functions. The Gaussian distribution is normally postulated, and a covariance function discussed in literature is chosen. The disadvantage of such a choice is that the probability measure can be concentrated on nonrelevant functions. The selection of a probabilistic model adequate to the problem in question remains a challenge.

A concept similar to Bayesian non-convex continuous optimization is used rather broadly, e.g. for solving problems of the multi-armed bandit. The decision space of the classical bandit problems is discrete, while it is continuous in the considered case. Nevertheless, some methods developed in the bandit theory are extended to the continuous decision space and are applicable to the continuous non-convex problems [35, 116]. The performance of algorithms using the bandit theory is characterized by a regret. The simple regret in the terms of non-convex optimization means the approximation error of the global minimum. Most results on convergence are valid for the objective functions which are elements of a reproducing kernel Hilbert space (RKHS) related to the covariance function of the underlying Gaussian process [11, 84, 126, 127, 132]. We will not discuss the theoretical details of that generalization as they are not directly related to our main questions which are addressed in the next sections. All the more so as such a discussion would require an in-depth analysis of the properties of RKHS and their interpretation.

2.2.1　Algorithms for Problems with Box Constraints

Most publications on Bayesian GO focus on algorithms for problems with simple constraints. We have been paying close attention to this topic since the difficulties in solving non-convex optimization problems with simple constraints are repeated in the case of more complex constraints in an even more unfavourable form. The original Bayesian algorithms are implemented directly according to the formulas (2.1.5) and (2.1.6). In the sequel, we will call such an implementation *the standard implementation*. Various recently developed versions of the standard implementations of Bayesian GO algorithms differ in underlying stochastic models and methods of solving the inner optimization problem (2.1.5); see, e.g. [53, 61, 75, 82, 88, 91, 106]. However, the presented results of numerical experiments are difficult to summarize as well as the recommended method for the solution of the inner optimization problem. Indeed, the most appropriate method is difficult to propose in part since the computational complexity of the problem increases with the number of iteration.

Several GO algorithms were developed recently using the Thompson sampling. This approach has proven successful in addressing the multi-armed bandit problem [1]. The classical Thompson sampling is applicable for discrete decision alternatives. Therefore a generalization is required for its application to select a current point of computation of the objective function in a continuous set. Thompson sampling has been used among others in [8] to reduce the local bias of search of MEI. The substitution of $y_{o,n-1}$ in (2.1.5) with an estimate of the minimum of the objective function $y_n^* < y_{o,n-1}$ can increase the globality of search. Such a MEI modification is developed in [8] using a Thompson sampling to obtain y_n^*. A more straightforward application of the Thompson sampling would be the generation of random points for the computation of the objective function according to the posterior distribution of the minimizer of the underlying GRF. Since such distribution for the stochastic functions applicable as models in GO are not known, different approximations are proposed in [9, 52, 58, 136] and references there.

The Upper Confidence Bound (UCB) is used as the acquisition for the development of Bayesian GO algorithms in several recent publications, e.g. [4, 59, 116]. The point for the computation of the objective function at iteration n of the UCB minimization algorithm is defined as follows:

$$x_n = \arg\max_{x\in\mathbf{A}} \left(-\mu_{n-1}(x) + \beta_n\sigma_{n-1}(x)\right), \tag{2.2.1}$$

where $\mu_{n-1}(\cdot)$ and $\sigma_{n-1}^2(\cdot)$ denote the conditional mean and conditional variance of a value of the GRF given data available at iteration n and β_n denotes an appropriate constant which defines the confidence level. This acquisition function has the properties of expected utility stated in Theorem 2.1 but does not satisfy the requirement (2.3.5) which excludes replicated observations at the currently best point; see Sect. 2.3. Although it is a theoretical disadvantage, the selection of model parameters implying replicated observations is not likely in practice. On the other hand, the convergence of the UCB maximization-based algorithm is shown for the functions of the related RKHS; the superlinear convergence rate in terms of regret is also proved in [116].

The information theory-based GO is quite similar to Bayesian GO. The original information theoretic algorithm was proposed in [119]; for the further results of this author and his collaborators, we refer to [120, 121]. These results are well presented in English in [121] where further references can be found. Several recently proposed algorithms are based on the same idea but use different probabilistic models of objective functions; see, e.g. [49, 50, 112, 128]. A current point for computing the objective function value by such algorithms is determined by the maximization of information about the minimum point. However, the maximization of information about the minimum is advantageous; computation of this criterion avoids many approximations in implementation, much of computational overhead and the limitations in choice of a probabilistic model [132]. The current point of the latter algorithm is defined by

$$x_{n+1} = \arg\max_{x \in \mathbf{A}} \left(H[p_x(y|x_i, y_i, \ i=1,\ldots,n)] \right. -$$

$$\mathbb{E}_{p(\eta|x_i, y_i, \ i=1,\ldots,n)} \left\{ H[p_x(y|x_i, y_i, \ i=1,\ldots,n,\eta)] \right\} \right), \qquad (2.2.2)$$

where $H(\cdot)$ denotes entropy, η denotes a hyper-parameter which mimics a minimum of the model function and $p_x(y|\cdot)$ denotes the conditional distribution density of the underlying GRF at point x.

The rationality of one-step optimality of Bayesian GO algorithms is questionable. Only few publications examine the multi-step optimal algorithms due to mathematical and numerical challenges. Progress has been made in [35] by means of a discretization of the feasible region, which is assumed to be a cubic mesh $Z = \{z_1, \ldots, z_M\}$, $z_i \in \mathbb{R}^d$. The minimization problem $\min_{x \in Z} f(x)$, $Z = \{z_1, \ldots, z_M\}$ has been considered where the objective function is modelled by a GRF $\xi(x)$, and the function values are observed in the presence of the additive uncorrelated Gaussian noise. Let $\tilde{x}_{on}(\mathscr{A})$ be an approximation of the global minimizer found by the algorithm $\mathscr{A}$ after n computations of the objective function values. Let N be a budget of computations of the objective function fixed in advance. The average value of the estimate of a global minimum $\mathbb{E}\{\xi(\tilde{x}_{oN}(\mathscr{A}))\}$ is accepted as the efficiency criterion of $\mathscr{A}$, and the optimal algorithm $\hat{\mathscr{A}}$ is defined by

$$\hat{\mathscr{A}} = \arg\min_{\mathscr{A}} \mathbb{E}\{\xi(x_{oN}(\mathscr{A}))\}. \qquad (2.2.3)$$

The optimal algorithm (2.2.3) is defined in [35] for a class of problems which are different from the problems considered in (2.1.2) in two assumptions: the feasible region Z is assumed discrete, and the noisy function values are assumed in (2.2.3). The theoretical analysis of (2.2.3) by the method of dynamic programming is easier than that of (2.1.2) because of discreteness of the feasible region. Nevertheless, the complexity of algorithm (2.2.3) remains too high for practical implementation. The challenges of multi-step optimal Bayesian optimization are considered also in

[42, 66, 88]. The results of analysis of a multi-step optimal algorithm for univariate optimization is presented in [24].

A one-step optimal algorithm is proposed in [35] as an approximation of (2.2.3) and named Knowledge Gradient. Possible versions of one step optimality are commented when defining algorithm (2.1.5). The Knowledge Gradient algorithm computes the objective function at n iteration at the point

$$x_n = \arg\min_{x \in \mathbf{Z}} \mathbb{E}\{v(x_1, \ldots, x_{n-1}, x, y_1, \ldots, y_{n-1}, \xi(x)))|$$

$$x_j, y_j, \ j = 1, \ldots, n-1\}, \qquad (2.2.4)$$

where $v(\cdot)$ is defined by (2.1.3). This algorithm is different from MEI defined by (2.1.5) since the former maximizes the conditional expectation of $\tilde{y}_{on}$ and the latter maximizes the conditional expectation of y_{on}.

The application of the original Knowledge Gradient method for high-dimensional problems can be complicated since the complexity of its iteration is $O(M^2 \log(M))$, $M = m^d$ where d is the dimensionality of the optimization region and m is defined by the requested accuracy [35]. For example, the value $m = 100$ would be set in case of the requested relative accuracy in each coordinate equal to 1% of the edge length of the feasible hyper-rectangle. Note that the objective function of the minimization problems (2.2.4) depends on conditional mean and variance of $\xi(x)$. Thus, their computational complexity at n-th iteration grows with n as n^3 if a standard algorithm is used for the inversion of $n \times n$ covariance matrix. The recent modifications of the Knowledge Gradient method were developed to a continuous feasible region and allowed using gradients and parallel observations of the objective function [139]. The modification for a continuous feasible domain avoided the discretization errors and explicit exponential complexity with respect to d. However, every iteration of this modification includes solving a two-level non-convex optimization problem. Further details of the Knowledge Gradient methods which include derivatives of objective functions can be found in [138, 140].

Recently several new ideas are proposed for the solution of the inner optimization problem. The solution of the highly multimodal inner optimization problem is computational intensive. A cross-entropy-based importance sampling method to cope with multimodality of those problems is proposed in [6]. An enhancement of the Monte Carlo method is considered in [137] for a general class of inner problems with any expected utility function as the objective, i.e. with the average of any utility function with respect to the conditional distribution of the underlying probabilistic model. The improvement probability and average improvement clearly satisfy the latter requirement. The Monte Carlo method presented in [137] provides an efficient evaluation of gradients of the expected utility functions. An important subclass of the objective functions of inner problem (including improvement probability and expected improvement) is discussed in [137] having the property of submodularity. Thus, greedy algorithms are applicable to solve the inner problem since the submodularity ensures finding a suboptimal solution.

The structure of the inner optimization problem implies its computational intensity, and finding a near-optimal solution takes a lot of computer time. The development of special sophisticated algorithms could reduce that computational intensity. On the other hand, it is not obviously reasonable to pursue the high accuracy of solving the inner optimization problem which validity is questionable due to subjectivity in selecting a stochastic model of objective function (covariance kernel and its parameters) and utility function. We discuss these problems in more detail in Sects. 2.3 and 2.5.

Parallel computing is a technology aiding the solution of computationally intensive problems. Parallel versions of Bayesian GO algorithms are described in [41, 44, 52, 58, 121, 131]. Although parallel algorithms are based on sequential prototypes, non-trivial modifications are usually required.

The publications rarely contain detailed reasoning to support the selection of a probabilistic model; its parameters usually are estimated by the maximum likelihood method. The focus is mainly on the solution of the inner optimization problem. Different methods have been applied, e.g. random search, evolutionary methods, DIRECT, ad hoc methods, etc. The performance of the applied method normally is illustrated by results of numerical experiments. However, it is difficult to draw a general conclusion about those experiments. The solution of the intractable inner optimization problem is the crucial challenge for standardly implemented Bayesian algorithms. That complexity limits the maximum number of the objective function evaluations. Although the budget of several hundreds can be sufficient for some applied problems with expensive objective functions, it limits the dimensionality of the expectedly solvable problems. The reliable solution of multimodal problems of dimensionality higher than 10 with such a budget does not seem realistic. This challenge limits also the application of the reviewed below Bayesian algorithms which are aimed at optimization problems with more general constraints, noise and available derivatives of the objective functions.

2.2.2 Algorithms for Problems with Complicated Constraints

The real-world optimization problems usually are with more complicated constraints than box constraints. Thus, Bayesian GO researchers pay lately more close attention to problems having general constraints

$$\min_{x \in A} f(x), \ \mathbf{A} = \{x : \ g_i(x) \le 0, \ i = 1, \ldots, m, \ h_i(x) = 0, \ i = 1, \ldots, p\}, \ x \in \mathbb{R}^d\},$$

(2.2.5)

where $g_i(x)$ and $h_i(x)$ can be linear as well as nonlinear functions. For many applied problems with expensive black-box objectives, $g_i(x)$ and $h_i(x)$ are defined either by analytical formulas or by simple algorithms which define geometric, economic, managerial, etc. conditions (see, e.g. [87, 168]). Bayesian GO algorithms with simple constraints can be quite easily extended to problems with constraints, which are not black-box expensive ones. It is sufficient to define a current point for the computation of $f(x)$ maximizing the expected utility/acquisition function over $\mathbf{A}$. Practically, however, the complexity added by the complicated constraints can drastically increase the computational intensity of solution of the inherent multimodal optimization problem.

Bayesian algorithms can be developed for problems where both, objective function and constraints, are defined by expensive black-box functions. Similar stochastic models are applicable for both counterparts of such optimization problems. However, to our knowledge, a theoretical generalization of the optimal Bayesian algorithm [78] to problems with stochastic constraints is not published anywhere. Possibly, this kind of theoretical analysis is not very interesting due to the obvious intractability of the problem.

Most known versions of Bayesian algorithms for constrained problems are modifications of MEI and other one-step optimal algorithms with simple constraints. Several types of black-box constraints can be identified and addressed differently. Explicit constraints are defined by (2.2.5) where the left hand functions are defined and continuous over some hyper-rectangle containing $\mathbf{A}$. Implicit constraints are defined by algorithms with binary output which indicates the infeasibility of an input, but does not evaluate its distance to the feasible region. The objective function can be either defined or not defined outside of the feasible region. Different cases are dealt with in different ways aiming at best performance of the developed algorithm.

Consider (2.2.5) with only inequalities. Let GRFs $\zeta_i(x)$ be accepted as a model of the black-box functions $g_i(x)$, $i = 1, \ldots, m$. The probability of the constraints violation at point x is evaluated by the conditional probability $P_n(x) = \mathbf{P}\{\zeta_i(x) \leq 0, i = 1, \ldots, m\}$ which depends at iteration n on the values of $g_i(x)$ computed at previous iterations. A natural extension of MEI would be the replacement of the expected improvement criterion by its aggregation with $P_n(x)$; see, e.g. [106, 107].

Methods of Lagrange multipliers are popular methods in mathematical programming. The possibility of their extension to black-box constrained problems using the augmented Lagrange function is addressed in [93]. Let $f(x)$, $g_i(x)$, $h_i(x)$ in (2.2.5) be expensive black-box functions. First, the inequality constraints are replaced by the equality constraints using slack variables:

$$h_{i+p}(x, s) = g_i(x) + s_{i+p}, \quad i = 1, \ldots, m, \quad s_i \geq 0.$$

Such a reformulation of the problem increases the number of variables. The authors claim, however, that it facilitates the solution. The Lagrange function of the reformulated problem is defined by the following formula:

$$L(x, s, \lambda, \rho) = f(x) + \lambda^T h(x, s) + \frac{1}{2\rho} \sum_{i=1}^{m+p} h_i(x, s)^2,$$

where $\lambda = (\lambda_1, \ldots, \lambda_{m+p})^T$, $s = (0, \ldots, 0, s_{p+1}, \ldots, s_{p+m})^T$, $(\lambda_{p+1}, \ldots, \lambda_{m+p})^T \geq 0$ and $\rho > 0$. The solution of the initial problem is reduced to a sequence of unconstrained optimization problems. A Bayesian algorithm for problems with box constraints is applied to minimize the Lagrangian which probabilistic model is composed of the models of $f(x)$, $g_i(x)$, $h_i(x)$; the minimization region is a hyper-rectangle $\hat{\mathbf{A}} \supset \mathbf{A}$. The authors of [93] claim that their algorithm *outperforms modern alternatives in several challenging constrained optimization problems*.

Various extensions of MEI are most popular Bayesian constrained optimization algorithms. The unconstrained algorithms based on other utility/acquisition functions have been similarly extended for the constrained problems. For example, an entropy reduction-based optimization algorithm for problems with implicit constraints is presented in [50, 51].

The minimizer of a constraint problem is frequently attained at the boundary of the feasible region, but the objective function is defined outside of that region. The computation of the function at an infeasible point is usually less informative than that at a feasible point. Nevertheless, the function values at infeasible points are important for defining next iterations. The proper handling of this kind of the asymmetry can enhance the search efficiency as shown in [71].

An interesting class of black-box GO problems are problems with implicit (also called unknown) constraints. The methods in [5, 38, 39, 43] are developed using a latent GRF for modelling feasibility under such constraints. For example, the sign of value of that latent GRF at point x is used in [5] as an indicator of x feasibility. Thus, two probabilistic models are involved to modify the expected improvement criterion by its hybridization with the probability of feasibility. The computational intensity of the modified criterion depends on the computational complexity of the joint conditional distributions of the involved GRFs. Original methods cope with these computational challenges. Nevertheless, the dependence of the complexity of the computation at iteration n depending on n and d requires further investigation.

The computational intensity related to implicit constraints can be reduced exploiting soft computing. The boundaries of A of the problems with implicit constraints can be approximated by known methods of pattern recognition. However, the requirements for a classifier used as a counterpart of the optimization algorithm are rather different from those in usual problems of pattern recognition: precise separation is required only in the neighbourhood of a global minimizer, and the training set of the classifier is rather small. Several algorithms hybridizing MEI with support vector machine classifiers are proposed in [7, 23, 103]. The results of several numerical examples are provided to illustrate the advantages of these implementations over the implementation based entirely on stochastic models.

Some of methods discussed in this subsection, e.g. [39, 43, 68], are applicable for noisy problems.

2.2.3 Noisy Optimization

The first publications on Bayesian GO, although not using this name, was on global optimization in the presence of noise [63, 64]. A one-dimensional algorithm is described which samples the objective function values at the points of maximum probable improvement. The probabilistic model of the objective function consists of a Wiener process, and the observations are corrupted by additive independent Gaussian noise. The formal similarity between Bayesian algorithms for noisy and noiseless problems is obvious from the very beginning of research in Bayesian GO. However, algorithms for noisy problems are more complicated from the implementation point of view. The increased complexity is caused by the much larger number of samples of noisy objective function values needed for reliable discovery of the global minimum, especially in the case of large number of local minima. The standard implemented algorithm could be inappropriate for noisy problems due to its inherent complexity where the time needed to compute a current trial point drastically grows with the number of iteration.

An exception is a special case of the Wiener process, for which special algorithms are presented in [18, 63, 151, 152] circumventing the explicit inversion of the covariance matrix. The experimental results in [155] have shown that global minima of highly oscillating functions are reliably found by this algorithm in the presence of hard noise; the time of inherent computations needed to perform several thousands of iterations is practically negligible. A similar method of avoiding the challenges of the inversion of covariance matrices of GRF ($d \geq 2$) is not known, at least to the authors of this book.

Replicated sampling of the function value at the same point can mitigate the influence of noise ensuring good performance in the one-variable case [152]. The multi-variable case is addressed in [54] where various replication strategies are compared in the presence of heterogeneous noise. The algorithm has been developed for a discretized feasible region; the cubic $10 \times d$ grid is used. Two testing examples of dimensionality $d = 2$ and $d = 6$ are solved. Several versions of the Bayesian algorithm based on different utility functions are compared; for all versions the initial sample is 55 replications at $10 \times d$ points which are randomly distributed in the feasible region. This data are used to identify the statistical model. The tested algorithms performed optimization with *low and high* budgets meaning 10 and 50 points for sampling the 55 replicas of function values. The results of experiments, however, are not sufficient to make clear conclusions, especially in case of heavy noise.

Various versions of Bayesian algorithms for noisy optimization are proposed in [62, 88, 92, 94] and others. The main difficulty is caused by the computational intensity of processing a large number of function evaluations needed for the noise filtration. The discretization of the feasible region to replicate sampling is a solution but only for low dimensions. The limited number of iterations, which is caused by the inherent complexity of Bayesian algorithms, severely narrows down the class of reliably solvable problems with respect to dimensionality, noise intensity and multimodality of objective functions. The research of noisy optimization is most challenging in Bayesian GO.

2.2.4   Methods Using Derivatives

Most Bayesian algorithms are derivative-free. The contemporary focus of Bayesian approach to derivative-free optimization is usually explained by the expensiveness of derivatives of the objective functions. Nevertheless, the potential advantage of using derivatives can be of interest for some applied problems where derivatives can be computed additionally to a function value at low additional cost. Technically, derivatives can be taken into account for a GRF rather simply, since their probability distribution is Gaussian. The covariance between a GRF values and derivatives are equal to the corresponding derivatives of covariance function of the GRF [143]. However, taking derivatives into account increases the complexity of the inner optimization problem since at iteration n a $(d + 1)n \times (d + 1)n$ covariance matrix should be processed instead of a $n \times n$ covariance matrix.

For example, let $\xi(x)$ be a stationary Gaussian stochastic process where $d=1$, $\mathbb{E}\{\xi(x)\} = 0$, $\mathbb{D}\{\xi(x)\} = 1$ and $r(t) = \exp(-(t/c)^2)$ is the covariance function. We assume that the process value and derivative are observed at point $x = 0$ and they are equal to y and y', respectively. The conditional mean $m(x|y, y')$ and conditional variance $s^2(x|y, y')$ are expressed by the known formulas related to multidimensional Gaussian distribution. The covariance between $\xi(0)$ and $\xi(x)$ is equal to $\exp(-(x/c)^2)$, the covariance between $\xi'(0)$ and $\xi(x)$ is equal to $\frac{2x}{c^2} \exp(-(x/c)^2)$, and the variance of $\xi'(0)$ is equal to $2/c^2$. Substitution of these formulas in the expression of Gaussian conditional mean and variance implies the following expressions:

$$m(x|y, y') = (y + x \cdot y') \exp(-(x/c)^2)$$

$$s^2(x|y, y') = 1 - \left(1 + \frac{2x^2}{c^2}\right) \exp(-2(x/c)^2).$$

The influence of the information on the derivative to the conditional mean and variance is illustrated in Fig. 2.1.

The algorithms using gradients are most efficient among algorithms for local minimization of differentiable functions. Although the gradient-based linear approximation is appropriate in the neighbourhood of the reference point, its accuracy deteriorates with distance from this point. Thus, the use of gradients in Bayesian GO is not necessary advantageous, especially since one gradient is equivalent to d function evaluations from the point of view of the computational complexity of the conditional distribution of the underlying GRF. The theoretical comparison of informativeness of a gradient with d function values is difficult. The numerical experiments started, to the best of our knowledge, in the 1980s of the last century [74, 159]. However, a reasonable testing was prevented due to the insufficient at that time computing power to invert large ill-conditioned covariance matrices.

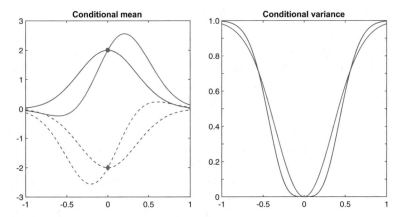

Fig. 2.1 Conditional means and variances presented in red and blue for the cases with and without derivative correspondingly; $y = 2$ (circle), $y = -1$ (diamond), and $c = 0.5$, $y' = 5$

A univariate algorithm using derivatives is theoretically and experimentally investigated in [17]. An unexpected conclusion about this algorithm is that the influence of derivatives to the search process vanishes when the number of iterations increases. Thus, the evaluation of the convergence rate of the multidimensional Bayesian algorithms using derivatives is particularly interesting.

Implementation aspects of algorithms with derivatives are discussed in [88, 139]. Various simplifications are proposed to cope with the difficulties of the inversion of large ill-conditioned matrices. In [138] a rescaling method is presented which enables circumventing the problem of ill-conditioning. The information on the directional derivatives is used in a version of the Knowledge Gradient method [140]. The computational burden implied by large covariance matrices is tackled in [32] by a dimensionality-reducing technique. The partition-based method with gradients is proposed in [158].

The published results are interesting, but the questions remain. Derivatives are definitely valuable for local approximation of a smooth function, but the rationality of their use for Bayesian global search is not entirely clear. The size of a matrix of covariances between $d + 1$ values of $\xi(x)$ is the same as the size of the matrix of covariances between a single value and d components of the gradient. These two alternatives should be compared with respect to the expense of function values and derivatives, computational intensity of the inner optimization problem and performance of the corresponding algorithms. A promising application area of the algorithms using derivatives is the optimization problems, where the derivatives are computed additionally to the function value with little additional effort. For example, such algorithms are of interest in relation to the prospective Infinity Computer where derivatives can be computed in addition to the function value for almost no extra computer time [110].

2.2.5 Extensions to Large Dimensions

Despite advantages of the standard Bayesian GO algorithms, their application is limited due to the relatively low dimensionality of the expectedly solvable problems. Increasing dimensionality of solvable problems is of great theoretical and practical interest. A crucial challenge is the limited number of objective function values which can be processed by a standard implemented algorithm. The challenges are caused by the computational intensity of the inner optimization problem as discussed in the previous sections. For example, the authors of [99] agree with our view on dimensionality of Bayesian GO: *it has a serious limitation when the parameter space is high-dimensional.* They confirm also that the main challenge is the complexity of the inner optimization problem. Since the traditionally used methods, such as DIRECT and multistart gradient, are not efficient, several potentially appropriate methods for solving the high-dimensional inner optimization problem are developed.

An experiment is described in [99] where the efficiency of a gradient method is enhanced by the appropriate scaling of the inner optimization problem. The performance of the presented method is illustrated by several examples up to dimensionality $d = 33$ with a budget $N = 150$ function evaluations. However, those examples are not sufficient to judge about the multiextremality and dimensionality of the optimization problems which could be reliably solved with a budget of 150 calculations. Anyway, quite large subsets of the 33 dimensional hypercubes are not explored regardless where the 150 computation points are located. Thus, it remains not clear what assumptions about the objective functions could guarantee that the global minimizer is not in such a subset with a large probability.

The computational intensity of the inner optimization problem is caused by the multimodality and the computational complexity of the utility/acquisition functions. The computation of the latter at iteration n includes the inversion of a $n \times n$ covariance matrix. Although computing time can be reduced by the development of special methods of matrix inversion, their complexity anyway will be $\Omega(n^2)$. Additional difficulties are implied by the ill conditioning of large covariance matrices. Recently several methods of lower complexity are developed for approximating the functionals of the conditional distributions needed for the implementation of GO algorithms; see, e.g. [30, 32, 33, 83, 130, 133] and references therein. Such methods seem interesting and deserve further investigation; however the classes of addressed problems should be defined more concretely to make predictions about the potential of these algorithms for high-dimensional Bayesian GO.

Decomposition is used for the development of algorithms for high-dimensional optimization problems of various types. Those algorithms exploit the favourable properties of the problem to be solved, e.g. the separability of the objective and constraints' functions. The separability assumption is favourable for the coordinate search type methods where low-dimensional Bayesian algorithms are applicable for the successive solution of subproblems. For example, the objective function is assumed additively separable with respect to the mutually exclusive lower

dimensional components in [99]. The Upper Confidence Bound method is applied
to the optimization of groups of variables where several strategies for the sequential
selection of these groups are used. A special case of separation, namely, the
separation of the effective and not effective variables, can be actual in high-
dimensional optimization. However, such a favourable property of a black-box
function is difficult to detect. The detection of the effective variables is especially
difficult where they are related to the original variables in a complex way.

Let us assume that the effective variables define a subspace of the original space
where the objective function is defined. The low-dimensional optimization method
can be applied to such a problem using a random linear mapping from the low-
dimensional search subspace into the original space. The method proposed in [135]
and named REMBO combines the mentioned random mapping with a Bayesian
algorithm. The results of several experiments are reported, including results of
minimization of the test functions of dimensionality $d = 25$ and $d = 10^9$ where
only two randomly chosen variables are effective. The test function values coincide
with the corresponding values of the well-known Branin test function defined for the
effective variables. The further results of theoretical and experimental investigation
of REMBO are presented in [134]. The method proposed in [26] also assumes that
the objective functions formally depend on many variables; however only few of
them are effective. The original idea of that algorithm is in including the active
learning into optimization process, i.e. in combining the optimization process with
the selection of variables. The projected-additivity assumption in [70] generalizes
the low effective dimensionality assumptions of the previously considered methods.
Correspondingly, the kernel of GRF, which is used for a probabilistic model of
objective functions, is assumed decomposable into a sum of kernels of lower dimensionality. The method proposed in [40] consists of decomposition and subsequent
optimization with respect to the groups of variables defined by the decomposition.
The decomposition is based on a sample of the objective function values at random
uniformly distributed points, and an UCB method is used for optimization. The
performance of the proposed method is illustrated by the solution of few synthetic
problems of dimensionality $d = 50$ and $d = 100$ performing up to 1000 samples
of the objective function values. A similar algorithm, called a dropout strategy,
optimizes a subset of variables at each iteration [69]. Although the discussed
experiments are interesting, the validity of the assumptions of separability and low
effective dimensionality for practical problems is difficult to check. Theoretical
as well as experimental assessment of the influence of the deviations from the
theoretical assumptions to the efficiency of the corresponding algorithms is difficult.
For example, the authors of [99] conclude that the proposed method *performs well
beyond the scope suggested by our theory*. Although such a conclusion sounds very
optimistic, a potential user would prefer a more concrete recommendation, albeit
a heuristic one, about recognition of problems for which the algorithm could be
recommended.

The extension of known Bayesian methods for high-dimensional problems is
an interesting topic. The search strategy, which is rational for the low-dimensional
problems, not necessary remains rational, when simply generalized for the high-

dimensional problems. In order to address the challenges of high dimensionality, it is foremost necessary to ensure efficient processing of significantly larger than before amounts of data accumulated during optimization. Despite various enhancements, the complexity of inner optimization problem limits the dimensionality of multimodal problems reliably solvable by the standard Bayesian GO methods. Some recently proposed methods are based on statistical models different from the standard in this approach Gaussian stochastic functions, e.g. random trees and forests. We do not discuss those methods since their theoretical properties have not yet been sufficiently investigated, and the published experimental results are rather fragmented.

2.2.6 Applications and What Publications Do Not Tell

The variety of applications of Bayesian GO algorithms is quite wide. A typical application area is optimal design of various technological plants and processes. Recently, the number of publications about applications in machine learning grew rapidly. The reports on Bayesian optimization algorithms to real-world problems emphasize, as can be expected, the expensiveness of the optimized black-box functions. The expensiveness the objective function of such applied problems is caused by the high computational intensity of the corresponding computer models. For example, computationally intensive models of fluid dynamics are used in the optimal design of aircraft wings [45, 47, 65, 97, 115, 144]. The models of chemical plants and processes include numerical methods of solution of systems of partial differential equations [31, 48, 101]. The complexity of computer models motivates the use of Bayesian optimization also in other areas, e.g. design of novel molecules [25, 46], economics [2], medicine [87] and various engineering optimization problems [29, 37, 73, 89]. Research in the field of machine learning is recently expanding rapidly, promoting also the use of Bayesian approaches in this field. Bayesian GO algorithms prove to be ideal for optimizing hyper-parameters of support vector machines [36], artificial neural networks [39, 60, 98, 114], kernel and logistic regression learning [140], extreme learning machine [67], random forest [141] and deep learning of block chains [77].

Providing a comprehensive review would not fit to the format of this book, and we did not have that purpose. We would only like to draw attention on one aspect of the presentation of experimental results. The optimized computer models are usually presented in detail; see, e.g. the cited above papers. However, the highly complex models do not highlight essential character of the objective function, which therefore is considered as a black-box. The choice of a Bayesian algorithm in that situation is well motivated. However, the substantiation of the choice of a concrete algorithm (kernel, acquisition function, method for estimating parameters, etc.) is complicated. The authors, seemingly, have used heuristic methods for eliciting information about the problem by discussing with the experts of the respective applied field. Nevertheless, that stage of problem solving as well as the influence of the human expertise in finding an acceptable solution is very rarely described in the papers.

The conclusion, that an acceptable solution is found, means either the fulfilment of a priori requirements or an improvement with respect to a known solution. The optimization results usually are well presented from the application point of view. Some properties of the objective function normally can be established by the analysis of data collected during optimization. For example, the visualization of data collected during optimization can be helpful for the post optimization analysis as shown, e.g. in [168]. To our opinion, reporting of such information would be desirable. The elicited information, as well as the heuristics about the multimodality of the optimized function, would be helpful for the further development of algorithms and recommendations for the potential users.

In some cases, the papers are completed only with a statement about the suitability of the used algorithm to find an acceptable solution, but no details are provided. Such a statement is not sufficient for a reader to evaluate the potential efficiency of the algorithm for his problem. For example, an optimal configuration of a wing-body of a modern civil transport aircraft is searched using the high-fidelity computational fluid dynamics simulation in [47]. An optimization problem of 80 variables is formulated and solved by a kriging-based method using 300 objective function evaluations. This is an excellent result from the engineering point of view. However, it is difficult to explain that success other than by the involvement of a human, since such a number of samples usually is not sufficient to find the global minimum of typical multimodal test functions of less than ten variables. A similar comment applies to the paper [55] where the optimal design of a multi-element airfoil is looked for. The optimal design problem is reduced to an optimization problem in a three-dimensional feasible region. First, 25 samples of the objective function at the points of orthogonal design are computed to construct a kriging model. A single computation of the objective functions is made by MEI which is based on the constructed model. The last computed function value is accepted as an approximation of the maximum. Since no other explanation is provided, we guess that the success can be explained by a simple, probably unimodal, hyper-surface of the objective function.

The authors of most papers on applications of the Bayesian GO express a positive opinion about the used methods. However the addition of relevant details would be desirable.

2.3   The Search Process as a Sequence of Rational Decisions

Algorithmic efficiency in computer science is evaluated most frequently in the worst-case setting. The analysis of the average efficiency is a reasonable alternative, especially where the worst case is unlikely in real life. Uniform distribution of points is the optimal algorithm for important classes of GO problems in the worst-case setting, and the sequential planning does not outperform the passive algorithm [122, 123, 157]. On the other hand, it is known from experience that the information elicited during search is helpful. Therefore, average optimal algorithms are especially interesting for GO theory. Bayesian algorithm defined by (2.1.2) is optimal with respect to the average approximation error of global minimum. However, the implementation of the optimal Bayesian algorithm consists of the numerical solution of the auxiliary problems which are intractable. The inherent high complexity of this algorithm complicates also its mathematical analysis. The substantiation of the selection of a GRF for an appropriate model is challenging as well as estimating its parameters; see Sect. 2.4. New ideas are needed to overcome the aforementioned challenges without abandoning the idea to develop an algorithm based on the theory of rational decision-making under uncertainty. The main reason of the implementation difficulties of the optimal Bayesian algorithm is the requirement of optimality of the final result which is obtained after many subsequent dependent iterations. On the other hand, the rationality of the optimization of the multi-step decision process seem questionable in view of adequacy of the chosen probabilistic model and its possible update during optimization.

We will build a GO algorithm based on the theory of rational decision-making under uncertainty. Consider iteration $n + 1$ of the solution of the problem

$$\min_{x \in \mathbf{A}} f(x), \ \mathbf{A} \subset \mathbb{R}^d, \tag{2.3.1}$$

where $y_i = f(x_i)$, $i = 1, \ldots, n$, have been computed at previous iterations. The choice of a point $x_{n+1} \in \mathbf{A}$ is a decision with an uncertain result. Finding a point with a slightly lower function value than the current record $y_{on} = \min_{i=1,\ldots,n} y_i$ is quite likely in a vicinity of the currently best point. On the other hand, the potential location of a global minimizer in the unexplored subregions cannot be excluded. We are going to resolve the ambiguity in choice with the aid of the theory of rational decisions under uncertainty. Accordingly, a model of uncertainty and a utility function $u_n(\cdot)$ should be defined. A random variable is a standard model of an unknown value; such a model in the context of global optimization is substantiated in [153]. The probability distribution of the random variable $\xi(x)$, which is accepted for a model of $f(x)$, generally speaking, depends on x_i, y_i, $i = 1, \ldots, n$. We will define the expected utility of a choice of x_{n+1} as recommends the decision theory

$$x_{n+1} = \max_{x \in \mathbf{A}} \mathbb{E}\{u_n(\xi(x)) \,|\, x_i, y_i, \ i = 1, \ldots, n\}, \tag{2.3.2}$$

where the conditional mean of $u_n(\cdot)$ is defined with respect to the information accumulated in the previous iterations. Let $p(y|\mu, \sigma)$ be the probability density of $\xi(x)$, where the mean value $\mu = \mu(x, x_i, y_i, \ i = 1, \ldots, n)$, and the standard deviation $\sigma = \sigma(x, x_i, y_i, \ i = 1, \ldots, n)$ depends on the results of previous computations. We assume that

$$\int_{-\infty}^{y_{on}} p(y|\mu, \sigma)dy > 0, \; y_{on} = \min_{i=1,\dots,n} y_i;$$

otherwise y_{on} would be the minimum of $\xi(x)$ with probability 1. Formula (2.3.2) can be rewritten in the following form:

$$x_{n+1} = \max_{x \in A} \int_{-\infty}^{\infty} u_n(y) p(y|\mu, \sigma)dy. \tag{2.3.3}$$

We focus on monotonically decreasing utility functions, since a minimization problem is considered. Further assumptions are needed to specify a sequence of utility functions $u_n(\cdot)$. For example, the sequence of $u_n(\cdot)$ which defines the optimal Bayesian algorithm can be derived by applying a dynamic programming method to the solution of (2.1.2). However, as mentioned above, the complexity of this problem is too high. The utility functions which define the algorithms (2.1.5) and (2.1.6) have been repeatedly discussed in the literature; see, e.g. [146]. An important parameter of these utility functions is y_{on}.

The expected utility of the selection of x for the current computation $U(x)$ is defined by the probability distribution of $\xi(x)$ and the utility function $u(\cdot)$:

$$U(x) = \mathbb{E}\{u(\xi(x))|x_i, y_i, \; i = 1, \dots, n\}.$$

The function $U(x)$ depends on x via the parameters of the conditional probability distribution of $\xi(x)$ which in turn depends on x and x_i, y_i, $i = 1, \dots, n$. We assume that the utility function $u(y) \geq 0$ is nonincreasing and has a finite expectation with respect to the considered distributions. The explicit definition of a uniquely appropriate $u(\cdot)$ is difficult, but some of its properties can be derived postulating favourable properties of $U(\cdot)$. We will show that x_{n+1} can be selected conforming the desirable properties of expected utility without specifying a concrete $u(\cdot)$.

We introduce new notation to derive an explicit dependence of expected utility on the parameters of the relevant distribution

$$U(\mu, \sigma) = \int_{-\infty}^{\infty} u(y) p(y|\mu, \sigma)dy = \int_{-\infty}^{\infty} u(\sigma y + \mu) \pi(y)dy, \tag{2.3.4}$$

where $\pi(y) = p(y|0, 1)$.

The choice of the point x to be evaluated next would be obviously not rational if the value $f(x)$ would be a priori known. Therefore we require the fulfilment of the inequality

$$U(\mu_1, \sigma) > U(\mu_2, 0), \; \sigma > 0, \tag{2.3.5}$$

for any values of μ_1, μ_2.

Theorem 2.1 *The inequality (2.3.5) is satisfied for the utility functions*

$$u(y) = c, \ y \geq \tilde{y}, \tag{2.3.6}$$

$$u(y) > c, \ y < \tilde{y}, \tag{2.3.7}$$

where $\tilde{y} < y_{on}$, $y_{on} = \min_{i=1,\ldots,n} y_i,$.

Theorem 2.2 *Let the function* $u(y)$ *be monotonously decreasing and satisfies the assumptions of Theorem 2.1. Let the probability density* $p(y|\mu, \sigma)$ *be symmetric with respect to the mean value* $\mu > \tilde{y}$ *and decreasing with* $|y - \mu|$. *Let the integral (2.3.4) be finite. Then* $U(\mu, \sigma)$ *is a decreasing function of* μ *and an increasing function of* σ.

For the proofs we refer to [160]. Without loss of generality, it can be assumed that $c = 0$. The following probability densities satisfy the assumptions of Theorem 2.2: the normal probability density

$$p(y|\mu, \sigma) = \frac{1}{\sigma\sqrt{2\pi}} \exp\left\{-\frac{(y-\mu)^2}{2\sigma^2}\right\},$$

and the Laplace probability density

$$p(y|\mu, \sigma) = \frac{1}{\sigma\sqrt{2}} \exp\left\{-\frac{|y-\mu|\sqrt{2}}{\sigma}\right\}. \tag{2.3.8}$$

A current iteration of MEI is defined, independently of the number of iteration, using the utility function

$$u_n(y) = \max\{y_{on} - y, \ 0\}, \tag{2.3.9}$$

where the utility is either proportional to the improvement of the approximation of minimum or equal to zero. However, such a definition of utility does not fit well with the trade-off between exploration and exploitation at different phases of search.

The algorithm of maximum probable improvement (P-algorithm) is defined using the following utility function:

$$u_n(y) = \begin{cases} 1, & \text{if } y < y_{on} - \varepsilon_n, \\ 0, & \text{if } y \geq y_{on} - \varepsilon_n, \end{cases} \tag{2.3.10}$$

where the utility can be related to the iteration number, by choosing an appropriate $\varepsilon_n > 0$ depending on n. The search strategy of this algorithm can be controlled taking into account the anticipated number of iterations.

The utility functions (2.3.9) and (2.3.10) satisfy (2.3.6). Next we present the corresponding formulas of expected utility which explicitly illustrate the statement of the Theorem 2.2. Let $p(x|\mu, \sigma)$ be the Gaussian density with mean value μ and standard deviation σ. The expected utility corresponding to (2.3.10) is equal to

$$U(\mu(x|x_i, y_i), \sigma(x|x_i, y_i)) = \Pi \left(\frac{y_{on} - \mu(x|x_i, y_i) - \varepsilon_n}{\sigma(x|x_i, y_i)} \right),$$

where

$$\Pi(z) = \int_{-\infty}^{z} \frac{1}{\sqrt{2\pi}} \exp(-t^2/2) dt.$$

It is obvious that for $\varepsilon_n > y_{on} - \mu(x|x_i, y_i)$, $U(\mu(x|x_i, y_i), \sigma(x|x_i, y_i))$ is a decreasing function of $(\mu(x|x_i, y_i)$ and increasing function in $\sigma(x|x_i, y_i))$.

Expected improvement is the expected utility for the utility function (2.3.9). The following formula

$$U(\mu, \sigma) = (y_{on} - \mu) \cdot \Pi \left(\frac{y_{on} - \mu}{\sigma} \right) + \frac{\sigma}{2\pi} \exp \left(-\frac{(y_{on} - \mu)^2}{\sigma^2} \right),$$

is frequently used in the analysis of the various implementations of the MEI algorithm. The character of dependence of $U(\mu, \sigma)$ on μ and σ is not obvious from this formula. However, the following formula derived in [148]

$$U(\mu, \sigma) = \sigma \int_{-\infty}^{\frac{y_{on} - \mu}{\sigma}} \Pi(t) dt, \tag{2.3.11}$$

reveals that $U(\mu, \sigma)$ is an increasing function on σ and a decreasing function on μ for $y_{on} < \mu$.

Recently several algorithms are developed where a current iteration is defined maximizing various acquisition functions. For example, the acquisition function named Upper Confidence Level (UCL) is used in [59, 116]. The UCL acquisition function

$$\alpha_n(x) = -\mu(x|x_i, y_i) + \beta_n \sigma(x|x_i, y_i), \tag{2.3.12}$$

is oriented to a minimization problems. It has some properties analogous to the properties of the average utility stated by Theorem 2.2. Moreover, this acquisition function can be expressed by the formula (2.3.4) where $p(\cdot)$ is Laplace, and pseudo-utility function is defined as follows:

$$u(y) = -y + \beta_n |y - \mu(x |\cdot)|, \tag{2.3.13}$$

where $\mu(\cdot)$ is the parameter of (2.3.8) meaning expectation. The function $u(y)$ in (2.3.13) is not a utility function in terms of the decision theory since it depends on a parameter of the model of uncertainty. We call it pseudo-utility function because of its similarity to a utility function in the definition of the current step of the algorithm (2.3.3). The function $\alpha_n(x)$ is obviously decreasing in μ and increasing in σ. However it does not necessary satisfy (2.3.9) for some μ, σ. The latter property can cause, although with a small probability, iterative selection of the currently best point.

The conditional variance is used by several authors as an acquisition function; see, e.g. [96]. This acquisition function corresponds to the average utility with respect to Gaussian distribution and the pseudo-utility function

$$u_n(y) = (y - \mu_n(x \mid \cdot))^2. \tag{2.3.14}$$

The search strategy of the algorithm based on maximization of such an average utility is close to the worst-case optimal strategy: the computed function values are ignored, and a point which is maximum weighted distance from $x_1, \ldots, x_n$, is selected as x_{n+1}. This algorithm is recommended as suitable for hybridization in [96]. It is interesting to note that formula (2.3.11) presents MEI as a hybridization of the locally biased ($\varepsilon_n = 0$) P-algorithm with an algorithm based on the pseudo-utility function (2.3.14).

2.3.1 Bi-objective Selection

A current iteration of a Bayesian GO algorithm is defined by the maximization of the average utility which is defined through the selected stochastic model and an appropriate utility function

$$x_{k+1} = \max_{x \in \mathbf{A}} U(\mu(x|x_i, y_i), \sigma(x|x_i, y_i)), \tag{2.3.15}$$

where $\mu(\cdot)$ and $\sigma^2(\cdot)$ denote the conditional mean and conditional variance of the GRF selected for the stochastic model of the objective function. The expected utility depends on the maximization variable $x \in \mathbf{A}$ via parameters of the statistical model $\mu(\cdot)$ and $\sigma(\cdot)$. Therefore, we can replace the maximization with respect to x with the maximization with respect to the vector $(\mu, \sigma)^T$ which varies over the feasible set of parameters of the statistical model $(\mu(x|x_i, y_i), \sigma(x|x_i, y_i))^T, x \in \mathbf{A}$. Although such replacement of variables does not seem advantageous from the implementation point of view, it implies an idea of the essential reformulation of the problem. Theorem 2.2 states that $U(\mu, \sigma)$ is a decreasing function of μ and an increasing function of σ for general assumptions about $u(\cdot)$ and $p(\cdot)$. Because of that property, the maximizer x^* of the average utility defines a point of the Pareto front of the set

$$\{(\mu(x|x_i, y_i), \sigma(x|x_i, y_i))^T\}, \quad x \in \mathbf{A}. \tag{2.3.16}$$

The complicated choice of an appropriate utility function with subsequent maximization of the expected utility can be circumvented by a selection of a point on the defined above Pareto front. The location of a point on the Pareto front transparently defines the exploration/exploitation trade-off at the current iteration of the corresponding Bayesian algorithm. A non-convex bi-objective optimization method should be applied for the selection of an appropriate Pareto solution because the Pareto front of (2.3.16) normally is non-convex. For example, a current site for computing the objective function value can be defined by the minimization of $\mu(x, x_i, y_i, \ i = 1, \ldots, n)$ under the restriction $\sigma(x, x_i, y_i, \ i = 1, \ldots, n) \geq \sigma_+$:

$$\sigma_+ = \gamma \max_{x \in A} \sigma(x, x_i, y_i, \ i = 1, \ldots, n), \ 0 < \gamma \leq 1.$$

The exploration/exploitation trade-off of the search can be easily controlled by selecting γ.

2.3.2   Numerical Example

We visualize in this subsection the bi-objective selection of the point to be sampled. A two-dimensional minimization problem (2.3.1) is considered, where $d = 2$, $A = \{x : \ (0, 0)^T \leq x_{.j} \leq (1, 1)^T, \ j = 1, 2\}$. We assume that n objective function values are computed at previous iterations, and a point x_{n+1} should be selected.

Consider the details of the experiment. The re-scaled well-known Rastrigin function

$$\varphi(z) = z_{.1}^2 + z_{.2}^2 - \cos(18 z_{.1}) - \cos(18 z_{.2}), \ -1 \leq z_{.i} \leq 1,$$

$$f(x) = \varphi(2x - (1, 1)^T), \ 0 \leq x_{.j} \leq 1,$$

is used as an example objective function. The quadratic 101×101 grid $\tilde{A}$ is used for the discrete representation of A. The feasible optimization region is discretized to facilitate the visualization of the feasible objective region. We have assumed that $n = 15$. The objective function values are computed at the points $x_i, \ i = 1, \ldots, n$, which are generated randomly with uniform distribution over $\tilde{A}$. The values y_i are obtained by the normalization of $f(x_i)$: $\sum y_i = 0$, $\sum y_i^2 = 1$. The example data are presented in the Table 2.1.

We interpret the experimental data below using two probabilistic models and a statistical model of the objective function. Both probabilistic models are isotropic homogeneous Gaussian random fields $\xi(x)$ where $\mathbb{E}\xi(x) = 0$, $\mathbb{E}\xi^2(x) = 1$. The correlation function of the Model 1 is $\rho_1(x, y) = \exp(-\lambda_1 \cdot ||x - y||)$, and the correlation function of the Model 2 is $\rho_2(x, y) = \exp(-\lambda_2 \cdot ||x - y||^2)$. The statistical model, i.e. Model 3, is a family of Gaussian random variables $\zeta(x)$ [153]. The average and variance of $\zeta(x)$ are defined by the following formulas:

Table 2.1 Data for the numerical example

i	1	2	3	4	5	6	7	8
x_{i1}	0.03	0.56	0.88	0.67	0.19	0.37	0.46	0.98
x_{i2}	0.12	0.59	0.23	0.38	0.58	0.25	0.29	0.62
y_i	0.82	1.08	0.83	−1.27	0.64	0.69	−1.04	0.81
i	9	10	11	12	13	14	15	
x_{i1}	0.16	0.86	0.64	0.38	0.19	0.43	0.48	
x_{i2}	0.27	0.82	0.98	0.73	0.34	0.58	0.11	
y_i	−0.61	−1.34	0.02	0.46	−1.40	1.34	−1.04	

$$\mu(x, x_i, y_i, \ i = 1, \ldots, n) = \frac{\sum_{i=1}^{n} y_i \frac{1}{w(x,x_i)}}{\sum_{i=1}^{n} \frac{1}{w(x,x_i)}}, \tag{2.3.17}$$

$$\sigma^2(x, x_i, y_i, \ i = 1, \ldots, n) = \sigma_o^2 \min \|x - x_i\|^2, \tag{2.3.18}$$

where $w(x, x_i)$ is an appropriate distance-related function. Formula (2.3.17) with $w(x, x_i) = \|x - x_i\|^2$ defines the classic also called Shepard interpolant.

The parameters λ_i are estimated applying the method of maximum likelihood to the data of the Table 2.1. We have applied the method of maximum likelihood since it is used in most published works on Bayesian GO. The challenges of the theoretical substantiation and practical application of that method for the fixed-domain deterministic data of computer experiments is discussed in Sect. 2.4. This procedure leads to the estimates $\lambda_1 = 16.9$ and $\lambda_2 = 10.2$. The used weights in (2.3.17) are defined by the formula

$$w(x, x_i) = \|x - x_i\|^2 \ln(\|x - x_i\|^2 + 1).$$

The feasible objective region is the set $(\mu(x|\cdot), \sigma(x|\cdot))^T$, $x \in \tilde{\mathbf{A}}$ where $\mu(x|\cdot)$ and $\sigma(x|\cdot)$ are conditional mean and conditional standard deviation of the corresponding GRF $\xi(x)$. The feasible objective region for the statistical model is defined similarly by the formulas (2.3.17) and (2.3.18).

The feasible objective regions corresponding to both probabilistic models are presented in the Fig. 2.2 by the cyan points; the Pareto front of the feasible objective region is shown by the green circles. The Pareto front of the feasible objective region represents the manifold of rational solutions at the current search iteration. The shape of the Pareto front depends on data as well as on the model: as shown in Fig. 2.2 it looks smooth for Model 1 but discontinuous for Model 2. The abruptness of the Pareto front is frequent for the models with fast-decreasing correlation functions.

An appropriate trade-off between exploration (represented by $\sigma(\cdot)$) and exploitation (represented by $\mu(\cdot)$) can be reached depending, for example, on the number of remaining iterations. The solutions in the objective space selected by the MEI and

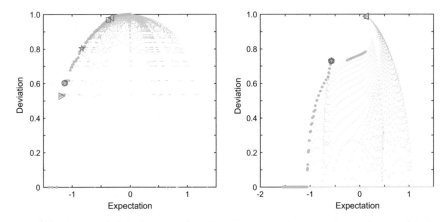

Fig. 2.2 The feasible objective region for a rational selection the next sample point; Pareto optimal front is indicated by green circles, and markers denote the special cases described in the text; the graph on the left presents data for Model 1, and the graph on the right presents data for Model 2

P-algorithm are shown in the Fig. 2.2. The blue circle marks the point selected by the MEI. A point selected by the P-algorithm depends on ε_n. Red markers, triangle to right, hexagram, pentagram, square and triangle to left, indicate the points which correspond to the following values of ε_{15}: 0.04, 0.4, 1, 3 and 5. An increase of ε_n leads to a change of the search strategy from locally biased to globally biased. The change is smooth for the smooth Pareto fronts but can be also discontinuous as illustrated by Fig. 2.2.

The influence of ε_n on the search strategy of the P-algorithm has also be discussed in the experimental and theoretical investigation presented in [146, 154]. Larger values of ε_n (exploration) are usually recommended for some starting iterations, while relatively small values are recommended when approaching the planned termination. However, the control of the exploration/exploitation trade-off by selecting an appropriate Pareto optimal solution is significantly more transparent than the selection of ε_n in the P-algorithm.

The points in the feasible optimization region which correspond to the points in the objective region are shown in Fig. 2.3 by the same markers as in Fig. 2.2. The blue rhombus marks the location of min y_i. The P-algorithm with some value of ε_n selects the same point as MEI; in such a case, both points normally are in a close vicinity of the best known point as illustrated by Fig. 2.3 where $\varepsilon_{15} = 0.4$.

There are not many cases where one can choose a GRF with desirable properties. The usual alternative is a choice from a limited list of the commonly used GRF. Next challenge is the estimation of parameters of a chosen GRF since the assumptions underlying the mathematically substantiated methods are not satisfied for the situations in question; see Sect. 2.4. Thus, a heuristic setting of the appropriate parameter values seems justified. Moreover, one can define a favourable shape of the Pareto front by selecting the appropriate parameter of the correlation function. For

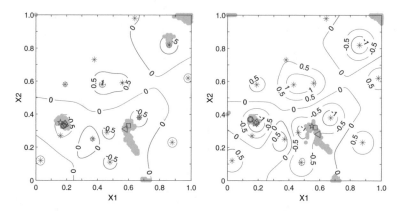

Fig. 2.3 Evaluated sample points are represented by stars, contours indicate the expected value of the objective function, the set of Pareto optimal decisions is presented by green circles, and the markers show the selected next sample point to be evaluated as described in the text; the graph on the left presents data for Model 1, and the graph on the right presents data for Model 2

example, the Pareto front of Model 2 seems very steep. The slope can be reduced by choosing a lower value of λ_2. Let Model 4 be the GRF with the exponential correlation function and $\lambda_2 = 3$. The results corresponding to the Model 4 are presented in Fig. 2.4.

Generally speaking, the candidate sampling points, presented in Figs. 2.3 and 2.4, are quite similar. This observation explains the long-known empirical result about the relative robustness of the Bayesian algorithm performance with respect to variation in the chosen GRF; see, e.g. [80]. This similarity extends to the so-called simple statistical models [153, 154]. Such statistical models are of interest to GO especially because they are of lower computational complexity than GRF. Pareto optimal solutions and decisions computed using the statistical model (2.3.17) and (2.3.18) are visualized in Fig. 2.5. The comparison of the visualized results suggest that using quite different models with appropriately chosen parameters can give similar results. Thus, the computational simplicity seems the most important criterion for choosing a probabilistic/statistical model of the objective functions.

The utility functions used, including the most popular (2.3.9), are difficult to validate. Thus, the design of GO algorithms based on bi-objective selection is advantageous since it does not require to choose a utility function. The trade-off between exploration and exploitation of the bi-objective selection-based search is easily and transparently controlled. The visualization of the search data in the bi-objective space well aids the interactive update of the parameters of the algorithm.

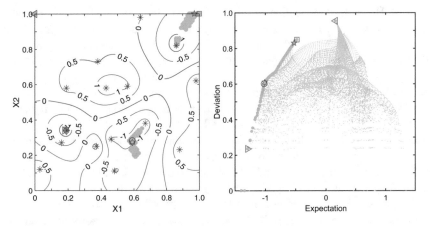

Fig. 2.4 Illustration of selecting next sample point by the considered algorithms (indicated by the markers defined in the text) using Model 4

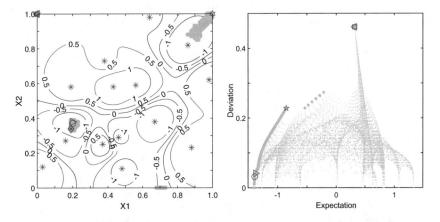

Fig. 2.5 Illustration of selecting next sample point by the considered algorithms (indicated by the markers defined in the text) using Model 3

2.4   Selection of a Probabilistic Model and Estimation of Its Parameters

2.4.1   Gaussian Random Fields

The design of an average optimal method is the original idea of Bayesian GO formulated in [78] where optimality means minimum average error with respect to a probability measure defined on a set of potential objective functions. From the point of view of applications, the average optimality can be senseless in case the probability measure is concentrated not on a set of functions of practical interest. However, constructive definition of a set of functions with certain desirable properties can be difficult. Even more difficult is the definition of a probability measure concentrated on a desirable set of functions. Besides compatibility between the probabilistic model and applied problems of interest, a selected model should be computationally tractable. The final step of the selection of an appropriate probabilistic model is the estimation of its parameters. Even in the case where a well-known random field is selected for a probabilistic model, the estimation of its parameters can be complicated due to the following reasons. The data used for estimation is a set of values of a deterministic function. The sampling of function values is allowed in the feasible optimization region which is bounded. Thus the estimators are applied under conditions which do not correspond to the assumptions of mathematical statistics implying known properties of estimates.

The feasibility of selection/construction of a provably appropriate probabilistic model is questionable and seldom discussed. The researchers normally choose a random field used in prior publications. GRF are solely considered, so only a covariance function is an option of choice. Frequently the objective function f is modelled by a homogeneous isotropic GRF $\xi(x)$, $x \in [0, 1]^d$, with a covariance function from the exponential class, i.e. assuming that $\mathbb{E}\xi(x) = \mu$ and

$$K_\gamma(x, z) = \mathrm{Cov}(\xi(x), \xi(z)) = \mathbb{E}(\xi(x) - \mu)(\xi(z) - \mu) =$$

$$= \sigma^2 \prod_{j=1}^{d} \exp\{-\lambda_j |x_j - x_j|^\gamma\} \, x \in \mathbb{R}^d, \, z \in \mathbb{R}^d. \quad (2.4.1)$$

The parameters of the model are $\mu \in \mathbb{R}$, $\sigma^2 \geq 0$, $\lambda_j > 0$ $(j = 1, \ldots, d)$ and $\gamma \in (0, 2]$. The model (2.4.1) is widely used not only in Bayesian GO but also in computer experiments; see, e.g. [104, 105]. The most influential parameter is γ; this is the reason why we use the notation K_γ for the covariance kernel.

The GRFs with covariance functions of the exponential class are frequently used in Bayesian GO for modelling objective functions; the following references show a variety of authors who used such covariance functions in their research [34, 57, 79, 113, 124]. The covariance functions of the Matérn class are popular in designing kriging models for the interpolation of spatial data [100, 117] and recently attracted attention also of researchers in Bayesian GO. The parameters of the chosen covariance function should be estimated using samples of values of the considered objective function. The maximum likelihood method is used most frequently for the estimation. However, the obtained result is not always applicable as shown in subsequent subsection.

2.4.2   Approximating K_2 by $K_{2-\varepsilon}$ with Small ε

The parameter γ determines the smoothness of realizations of the random field with covariance function (2.4.1). We distinguish three situations: (i) for $\gamma \in (0, 1]$ the realizations are not differentiable, (ii) for $\gamma \in (1, 2)$ the realizations are exactly once differentiable in each direction with the derivatives satisfying Hölder condition with parameter $\gamma - 1$, and (iii) for $\gamma = 2$ the realizations are differentiable infinitely many times. In view of this and since a typical objective function is quite smooth, it is tempting to assume $\gamma = 2$. This assumption, however, leads to difficult mathematical and computational problems. Assume $d = 1$, $\sigma^2 = 1$, $\lambda = 1$. Let us discretize the interval $[0, 1]$ and replace the random process $\xi(x)$, $x \in [0, 1]$ with an N-dimensional random vector $\xi_N = (\xi(x_1), \dots, \xi(x_N))^T$, where $x_i = (i - 0.5)/N \in [0, 1]$, $i = 1, \dots, N$. The mean of ξ_N is $\mu 1_N$, where $1_N = (1, 1, \dots, 1)^T$ is N-vector of ones; the covariance matrix of ξ_N is $W_N = (K_\gamma(i/N, j/N))_{i,j=1}^N$.

To simulate realizations of ξ_N or to implement any version of the Bayesian optimization algorithm, one needs to invert the covariance matrix W_N. Figure 2.6 demonstrates that numerical inversion of W_N could be extremely hard for $\gamma = 2$ and rather easy for $\gamma = 2 - \varepsilon$ even if $\varepsilon > 0$ is very small. For example, for $N = 100$ we have $\log_{10}(\lambda_{\min}(W_{100})) \simeq -268.58623$ if $\gamma = 2$ and $\log_{10}(\lambda_{\min}(W_{100})) \simeq -8.36979$ if $\gamma = 1.9999$. It is rather counter-intuitive, but interesting enough, the matrices W_N have completely different properties for $\gamma = 2$ and $\gamma = 1.9999$. On the other hand, in the range $0.5 < \gamma < 1.999$ the matrices W_N seem to have very similar properties. Summarizing this discussion, we claim that substituting K_2 with $K_{2-\varepsilon}$ with small ε is wrong for two reasons: (a) the corresponding covariance

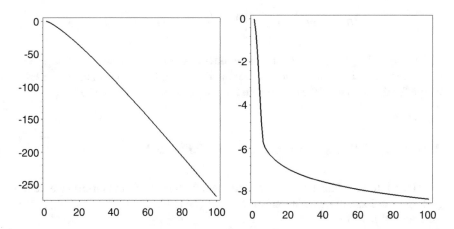

Fig. 2.6 Values of $\log_{10} \lambda_{\min}$, the decimal logarithm of the minimal eigenvalue, for the matrix W_N. Left: $\gamma = 2$ and $N = 1, \dots, 100$. Right: $\gamma = 1.9999$ and $N = 1, \dots, 100$

matrices W_N have completely different properties, and (b) the smoothness of realizations of the corresponding random processes or fields is completely different.

2.4.3 Inverting the Covariance Matrix W_N in the Case $\gamma = 2$

In this section, we provide explicit expressions for the elements of the inverse of the covariance matrix for the squared exponential covariance kernel K_2. Assume first that $d = 1$ and the points $x_1, \ldots, x_N$ are equally spaced on $[0, 1]$; for example, $x_i = (i - 0.5)/N$, $i = 1, \ldots, N$. In this case, the covariance matrix W_N is $W_N = (w^{(i-j)^2})_{i,j=1}^N$ with $w = e^{-1/(\lambda N^2)}$.

For two non-negative integers k and m and any real $q \neq 1$, the Gaussian binomial coefficient $\binom{k}{m}_q$ is defined by

$$\binom{k}{m}_q = \frac{(1 - q^k)(1 - q^{k-1}) \cdots (1 - q^{k-m+1})}{(1 - q)(1 - q^2) \cdots (1 - q^m)} \quad \text{for } 0 \leq m \leq k$$

and 0 otherwise. We shall also use the notation $C(q; i, j) = \sqrt{\prod_{k=i+1}^{j-1}(1 - q^{2k})}$. Using this notation, results of [72] imply that the matrix $W_N = (w^{(i-j)^2})_{i,j=1}^N$ has the Cholesky decomposition $W_N = LL^T$, where $L = (l_{ij})$ is lower triangular matrix with $l_{ij} = 0$ for $i < j$ and $l_{ij} = w^{(i-j)^2} C^2(w; i - j, j)/C(w; 0, j)$ for $i \geq j$. For the inverse of W_N, we have $W_N^{-1} = (L^{-1})^T L^{-1}$, where $L^{-1} = (\tilde{l}_{ij})$ with $\tilde{l}_{ij} = 0$ for $i < j$ and

$$\tilde{l}_{ij} = (-w)^{i-j} \binom{i-1}{j-1}_{w^2} \Big/ C(w; 0, i) \quad \text{for } i \geq j.$$

If $d > 1$ and the points x_i constitute a product grid, then the covariance matrix W_N is a Kronecker product of the covariance matrices in different coordinates; the same is true for the inverse of the covariance matrix; see details in [72], p.883.

2.4.4 MLE Estimation of μ and σ^2 for $\gamma = 2$

Consider f to be realization of a random field with mean μ and covariance kernel (2.4.1), $\gamma = 2$ and N observations of f have been made at points $x_1, \ldots, x_N$. MLE estimator of μ is $\hat{\mu} = F_N^T W_N^{-1} 1_N / 1_N^T W_N^{-1} 1_N$, where $F_N = (f(x_1), \ldots, f(x_N))^T$. The log-likelihood function of σ^2 (multiplied by 2) is

$$2L(\sigma^2) = - N \log 2\pi - N \log \sigma^2 - \log \det(W_N) - \frac{1}{\sigma^2}(F_N - \hat{\mu} 1_N)^T W_N^{-1}(F_N - \hat{\mu} 1_N)$$

and therefore the MLE of σ^2 is

$$\hat{\sigma}^2 = \frac{1}{N} \left(F_N^T W_N^{-1} F_N - \frac{(F_N^T W_N^{-1} 1_N)^2}{1_N^T W_N^{-1} 1_N} \right).$$

2.4.5 Numerical Experiments

The choice of an underlying probabilistic model can be critical to the performance of the corresponding Bayesian GO algorithm. A homogeneous isotropic GRF with quadratic exponential covariance function is an attractive model because of smoothness of its sampling functions. Once a model is selected, its parameters need to be evaluated. As mentioned above, estimating model parameters is a challenging problem. In this section, we present some numerical results illustrating these challenges. The formulas defining the maximum likelihood estimates of μ and σ^2 are presented above. A univariate optimization problem should be solved to estimate the scale parameter λ using the maximum likelihood method; the algorithm described in [150] or similar one is appropriate here. The application of simple-looking MLE formulas for $N > 50$ is challenging since the condition number of the correlation matrix normally is huge and the computation error remains unclear. Such an error can affect the performance of the corresponding Bayesian GO algorithm. On the other hand, even the theoretical analysis of properties of the estimates for deterministic data is difficult. Therefore, we performed an experiment to compare the theoretical (exactly computed) estimate for a given objective function with the routinely computed estimate. The theoretical estimates have been computed in our experiments using exact arithmetic of MATLAB (indeed using the arithmetic of rational numbers). The estimates for comparison have been also computed according to the same formulas using double precision arithmetic. The sites of observations x_i are uniformly distributed in the interval $[0, 1]$. We considered the cases $f(x_i) = x_i^p$, where $p = 0$ and $p = 1$, to maintain comparability with [142]. First, we have assumed that μ and λ are known: $\mu = 0$, $\lambda = 1$. To ensure computations in rational numbers, we had to approximate values of exponential function correspondingly. Let w be a rational approximation of $\exp(-1/N^2)$, implying rationality of correlations between observations at points i/N and j/N: $r_{i,j} = w^{(i-j)^2}$. The ML estimates of σ^2 based on N observations are given in Table 2.2. The estimates, obtained using exact computations, change with increasing n as proved theoretically in [142]: $\hat{\sigma}_n^2 \to 0$ for $p = 0$ and $\hat{\sigma}_n^2 \to \infty$ for $p = 1$. However, $\hat{\sigma}_n^2$, computed using double precision arithmetic, changes with increasing n differently than the theoretical estimate.

Next, we have assumed μ unknown and computed $\hat{\mu}$ and $\hat{\sigma}^2$ using the same data. For the case $p = 0$, the obvious results $\hat{\mu} = 0$, $\hat{\sigma}^2 = 0$, are quite natural. The results for the case $p = 1$ are presented in Table 2.3. In both cases $\hat{\mu}$ approaches the average of observations, but $\hat{\sigma}^2$ behaves similarly to the case of $\mu = 0$ (Table 2.2).

Table 2.2 Estimates of variance; $\mu = 0, \lambda = 1$

	$N=5$	$N=10$	$N=15$	$N=20$	$N=25$	$N=30$	$N=50$
$p=1$, Double prec.	0.3978	0.3967	0.3395	0.2547	0.2037	0.1697	0.1016
$p=1$, Exact	0.4200	0.5477	0.5660	0.6805	0.7021	0.7973	0.9930
$p=0$, Double prec.	0.3862	0.2511	0.1682	0.1263	0.1011	0.0843	0.0506
$p=0$, Exact	0.3851	0.2506	0.2117	0.1777	0.1622	0.1453	0.1127

Table 2.3 Estimates of mean and variance, $p = 1, \lambda = 1$

		$N = 5$	$N = 10$	$N = 15$	$N = 20$	$N = 25$	$N = 30$	$N = 50$
$\hat{\mu}$	Double prec.	0.6000	0.5500	0.5331	0.5249	0.5200	0.5167	0.5100
	Exact	0.6000	0.5500	0.5333	0.5250	0.5200	0.5167	0.5100
$\hat{\sigma}$	Double prec.	0.2588	0.3208	0.2917	0.2199	0.1764	0.1472	0.0885
	Exact	0.2813	0.4719	0.5064	0.6315	0.6582	0.7585	0.9637

Finally, we have considered the case of unknown λ. The data is defined by $f(x_i) = x_i$ for the uniformly distributed x_i, $i = 1, \ldots, 20$. We have used exact (rational number) arithmetic for maximization of logarithmic likelihood of λ. However, the estimate appeared infeasible. The increase of λ implied the monotonic increase of the likelihood function; $\hat{\sigma}^2$ is also monotonously increasing, and $\hat{\mu} = 0.525$ remained unchanging. For further examples we refer to [147].

The presented analysis and numerical examples highlight the challenges in selecting an appropriate GRF for a model of objective functions as well as in estimating its parameters.

2.5 Partition-Based Algorithms

2.5.1 Implementation

The inherent computational intensity of a current iteration of a standard implemented Bayesian GO algorithm increases rapidly with the number of iterations. A non-convex inner optimization problem should be solved at every iteration. This problem is especially challenging since the optimized function includes the inversion of the $n \times n$ covariance matrix , the condition number of which increases with the number of iteration n. Therefore, the number of practically executable iterations of a standard implemented Bayesian GO algorithm is quite limited. A possible way to reduce the computational burden is to implement multidimensional algorithms similar to the univariate algorithms by sequential partition of the feasible optimization region [148, 150]. This kind of implementation is natural since the used univariate statistical models are Markovian stochastic processes. The developed algorithms are efficient not only in terms of number of samples of the objective function needed to find the global minimum but also in terms of the complexity of inherent computations performed at a current iteration. Therefore, it is natural to extend the partition-based implementation to the multidimensional case. Another argument for such an extension is the successful partition-based implementations of algorithms in other branches of GO, e.g. in Lipschitzian GO [56, 95, 108, 109].

The experimentation started with the simplicial partition [156]. The algorithm is initialized by either covering or partitioning the feasible region by a predefined (small) number of simplices. At further iterations, a simplex is selected for partition according to the correspondingly adapted criterion of the improvement probability. The computational burden is avoided since the improvement probability p_S is defined conditionally to the function values at the vertices of the considered simplex S unlike to the standard implementation where all available function values are taken into account:

$$p_S = \max_{x \in S} \Pi \left(\frac{y_{on} - \mu(x | x_i, y_i, \ i \in I) - \varepsilon_n}{\sigma(x | x_i, y_i, \ i \in I)} \right), \tag{2.5.1}$$

where $\Pi(\cdot)$ denotes Gaussian cumulative distribution function; I is the set of indices defining the vertices of the simplex S; y_{on} is the best found so far objective function value; ε_n is the improvement threshold; and $\mu(\cdot|\cdot)$, $\sigma^2(\cdot|\cdot)$ are conditional mean and variance respectively. Further simplifications are based on the approximation of (2.5.1) by the analytical expressions proposed in [166, 167]. The selected simplex is partitioned, most frequently, by bisection.

Delaunay triangulation is applied in [19] to partition the feasible region by simplices as close as possible to regular ones. A new approximation of (2.5.1)

$$\frac{y_{on} - \frac{1}{|I|} \sum_{i \in I} y_i - \varepsilon_n}{V}, \tag{2.5.2}$$

follows from taking the monotonicity of $\Pi(\cdot)$ into account and replacing $\mu(\cdot|\cdot)$, $\sigma^2(\cdot|\cdot)$ by their asymptotic expressions [161]; V denotes the volume of the simplex. The convergence rate of this algorithm is established generalizing the previous results on convergence of univariate algorithms [146] to the multidimensional case. It is shown that the maximum error of evaluation of global minimum after large enough number of the objective function samples n is at most $c_1 \exp(-c_2 \sqrt{n})$ for c_1, c_2 depending on the dimension of the feasible region and on f. Despite the high asymptotic convergence rate and favourable testing results with low-dimensional test functions, that algorithm can be not attractive for high-dimensional problems, because the computational intensity of Delaunay triangulation grows rapidly with dimension. As an alternative, the rectangular partition for the design of the P-algorithm is implemented in [22]. A hyper-rectangle is selected for subdivision according to the criterion similar to (2.5.2). The selected hyper-rectangle is bisect by a hyperplane orthogonal to the longest edges of the hyper-rectangle. The objective function values are computed at 2^{d-1} intersection points at every iteration.

Such an algorithm is attractive for parallelization. The convergence rate of the rectangular partition-based algorithm is similar to that of the Delaunay partition-based algorithm [19].

The computational intensity of partition-based algorithms is relatively low since the criterion of selection of a hyper-simplex/hyper-rectangle for partition depends only on the function values at the vertices of the polytope. The use of local information in the evaluation of a polytope is justified since the conditional distribution of $\xi(x)$ basically depends on the closest to x observation points x_i and corresponding y_i. The globality of search is guaranteed by the selection of a simplex according to (2.5.1) and by maintaining the priority queue of polytopes based on the criterion (2.5.2). The priority queue is updated at every iteration ensuring that every polytope is eventually selected for the partition, and in this way the global convergence of the algorithm is ensured. The partition based Bayesian GO is generalized to multiobjective case and proved well suitable for the optimal design of biochemical plants [163, 168].

A bi-objective selection-based Bayesian algorithm is implemented in [163]. The simplicial partition is used. The non-dominated simplices are selected for subdivision where the expectation and deviation objectives are computed using the GRF model with squared-exponential covariance function. Two objectives are used to evaluate a subregion also in some other partition-based GO methods, e.g. in the algorithms using a Lipschitzian model [56, 90, 118]. However, the advantage of the bi-objective selection-based Bayesian algorithms is their theoretical substantiation by the theory of rational decision-making.

Partition-based derivative-free algorithms have advantages over standard implemented algorithms with respect to the inherent computational burden. Therefore, the partition-based implementation seems promising also for algorithms using derivatives. Consider the rectangular partition. The feasible region is partitioned at a current iteration into sub-rectangles. A rectangle is selected and partitioned into three parts by two hyperplanes which are rectangular to the longest edge and subdivide this edge into three equal subintervals. Such partition is used, e.g. in [56] where the selection criterion is based on a Lipschitzian model of the objective function. In the version of the P-algorithm proposed in [158], a hyper-rectangle for the partition is selected using information about the function value and gradient at the centre of the hyper-rectangle.

2.5.2 Convergence

Asymptotics of Bayesian GO algorithms attracted attention of researchers only recently seemingly because of the following reasons. Bayesian algorithms are aimed mainly at expensive black-box optimization problems. The number of iterations in such tasks is quite limited because of the long-running computation of the objective function values. However, the maximum number of iterations of standard implemented Bayesian algorithms is quite limited due to the computational complexity of the inner problem despite the expensiveness of the objective function. Note that the computational complexity increases in the iteration. Thus, convergence studies are not a priority for the researchers due the practically limited number of iterations. An exception is the single variable algorithms which are free from the challenges of the computational intensity. For the first results on the convergence of an univariate Bayesian GO algorithm, we refer to [148]. The investigation of convergence of other original Bayesian algorithms during the 1980s and 1990s was not active; for the results of this period, we refer to the monographs [79, 81, 125, 146]. The results can be summarized as follows: for a continuous function $f(x)$, $x \in \mathbf{A} \subset \mathbb{R}^d$,

$$\lim_{n \to \infty} y_{on} = \min_{x \in \mathbf{A}} f(x), \tag{2.5.3}$$

where $\mathbf{A}$ is a compact set, $y_{on} = \min_{i=1,\dots,n} y_i$, $y_i = f(x_i)$, and x_i is a point of the computation of the objective function at iteration i.

Recall that the inherent computational complexity of the partition-based algorithms is much lower than the complexity of standard implemented Bayesian algorithms. The maximum implementable number of iterations of the partition-based algorithms is considerably large. The increase in the maximum number of iterations stimulated interest in the asymptotic properties of partition-based algorithms.

Bayesian algorithms are based on probabilistic models, usually on GRFs. Although it is based on a probabilistic model, a Bayesian algorithm is deterministic, and it is used for the minimization of deterministic objective functions. Therefore, deterministic convergence is of interest. The questionable adequacy of the selected probabilistic model to the considered problem also motivates the investigation of the deterministic convergence. The natural weakest mathematical assumption about the objective functions of black-box problems is continuity. Starting from the first Bayesian algorithms, the convergence analysis is focused on continuous functions. The convergence of the one-step optimal Bayesian algorithm based on Wiener process model for continuous univariate functions is proved in [148]. A simple idea of this proof is to show that the sequence of points produced by the algorithm is almost everywhere dense in the feasible region. The generalization of this idea for the optimization of functions of several variables is not easy. Difficulties are caused by the complicated dependency of conditional mean and variance, $\mu_n(x, |x_i, y_i, i = 1, \dots, n)$, $\sigma_n^2(x, |x_i, y_i, i = 1, \dots, n)$, on $x_i, y_i, i = 1, \dots, n$, and on x. Markovian random processes are an exception since the conditional mean and conditional variance at a point x depend only on the neighbouring computed values of the objective function; this advantage is exploited in [148].

Recent results on convergence of Bayesian GO algorithms are proved for the objective functions which are elements of a reproducing kernel Hilbert space (RKHS) related to the covariance function of the underlying Gaussian process [16, 118, 182, 183, 196]. We do not discuss such type of convergence because of absence of results related to partition-based algorithms, and it would require an in-depth discussion of the properties of RKHS and their interpretation.

The convergence rate of an optimization algorithm is an important criterion of its efficiency. The studies of the convergence rate are started with single-variable case. For example, for various versions of the P-algorithm, the convergence rate $\Delta_n = O(n^{-2})$ is established in [17, 21] assuming that the objective function is twice differentiable with a positive second derivative at the global minimizer; Δ_n is an upper bound for the error of the approximation of the global minimum. The convergence rate of a special version of the P-algorithm is $\Delta_n = O(n^{-3+\delta})$, $\delta > 0$, as shown in [16].

One of the first partition-based multidimensional algorithms is a version of P-algorithm based on the simplicial partitioning [156]. Further modifications are proposed in [166, 167], and their convergence for continuous objective functions (in the sense of (2.5.3)) is proved. The convergence rate of P-algorithm based on simplicial and rectangular partition is evaluated in [19, 22]. We cite the theorem from [22] about the convergence rate of a rectangular partition-based algorithm.

Theorem 2.3 *Let $f(x)$, $x \in [0, 1]^d$ be twice continuously differentiable function that has a unique minimizer x^*. There is a number $n_0(f)$ such that for $n \geq n_0(f)$, the residual error after n iterations of the algorithm satisfies*

$$\Delta_n(f) \equiv \min_{1 \leq i \leq n} f(x_i) - f(x^*) \leq \frac{4^{2/d}}{8} \left\| D^2 f \right\|_{\infty, [0,1]^d} (q \cdot d) \exp\left(-\sqrt{n}\beta(f, d)\right),$$

$$\beta(f, d) = \left(\frac{2\sqrt{\pi} \left(\det(D^2 f(x^*))\right)^{1/2}}{9ed^{(d+3)/2}(\pi eq)^{d/2}} \exp\left(-\frac{1}{2} - \frac{16}{q}\right) \right)^{1/2}.$$

$$\left\| D^2 f \right\|_{\infty, K} \equiv \sup_{x \in K} \sup_{\substack{u_1, u_2 \in \mathbb{R}^d \\ \|u_i\|=1}} \left| D_{u_1} D_{u_2} f(x) \right|,$$

where $D_y f$ is the directional derivative of f in the direction y; $K \in [0, 1]^d$ is a compact set; $q \approx 1.27$.

In short, the theorem states that eventually the error is at most $c_1 \exp(-c_2\sqrt{n})$ for c_1, c_2 depending on d and f. Let us note that $n_0(f)$ after which the bound applies can be large since it depends (exponentially) on the seminorm $\left\| D^2 f \right\|_{\infty, [0,1]^d}$.

An idea to adapt the probabilistic model to the data acquired during optimization seemed attractive from the start of investigation in Bayesian GO. The potentially applicable methods for updating estimates of parameters of GRF are developed and investigated by the experts in statistics. Thus, the belief is quite natural that the adaptation of the probabilistic model can improve the efficiency of the respective GO algorithm. However, the investigation of adaptive algorithms, particularly of their convergence, can be challenging. To our best knowledge, the properties of an estimator of a parameter of GRF using dependent observations of a computer experiment are not investigated. Unlike the assumptions in the statistical investigation, our data are obtained by measuring values of a deterministic function. The difficulties in fixed-domain estimation of the parameters of a probabilistic model using the deterministic (computer model produced) data are discussed in Sect. 2.4 where some unfavourable results with respect to the common estimating praxis are presented. The following example highlights the discrepancy between the properties of an estimator with respect to the stochastic and deterministic model. Let's analyse the estimators of the parameter of the Wiener process σ^2. The statistical properties of the maximum likelihood estimator of σ^2 seem favourable for the adaptive updating in an optimization algorithm. Indeed, this estimator is unbiased and consistent for the observations of a Wiener process at any set of points scattered in the optimization interval; the sequentially selected points can also depend on values observed at previously selected points [165]. However, the maximum likelihood estimator $\hat{\sigma}^2$ applied for the values of a continuously differentiable function converges to zero. The Wiener process-based MEI algorithm is developed in [148] where its convergence is proved for any continuous objective function assuming any fixed $\hat{\sigma}^2 > 0$. The dependence of the efficiency of the algorithm on $\hat{\sigma}^2$ is investigated experimentally, and a recommendation for the selection of an appropriate parameter is provided. However, the convergence of the adaptive algorithm for differentiable functions using a vanishing parameter $\hat{\sigma}^2$ is questionable. The investigation of more general cases is extremely difficult.

The Bayesian approach is prone to the formal description of methods for optimization in the presence of noise. Two probabilistic models are involved in the definition of such algorithms: a model of objective functions and a model of noise. The former model normally is a GRF, and the latter is a set of independent Gaussian random variables. The convergence for a deterministic objective function is considered in the probabilistic sense with respect to the probabilistic model of noise. The first algorithm for noisy optimization is proposed by Kushner [63] a decade before the title *Bayesian global optimization* is introduced. The convergence of the proposed algorithm and the influence of the improvement threshold ε_n to its performance are discussed in [63]. However, neither the convergence proof nor testing results are provided.

The maximum average improvement algorithm for noisy optimization proposed in [152] is based on Wiener process model of the objective functions and on the assumption that the observation errors are independent Gaussian random variables. We will briefly comment the convergence of this algorithm. The feasible interval is discretized to facilitate the processing of large number of function evaluations which is needed to cope with noise: $\mathbf{A} = \{v_1, \ldots, v_N\}$, $v_i = \frac{i-1}{N-1}$. The objective function values are observable with the additive errors: $f(v_i) + \xi_i$ where ξ_i are independent Gaussian random variables with zero mean and fixed variance. The current approximation of $f_{min} = \min_{1 \leq i \leq N} f(v_i)$ is computed using the formula of the conditional mean of Wiener process observable in the presence of noise $m(\cdot | \cdot)$

$$y_{on} = \min_{1 \leq i \leq N} m(v_i | (x_j, y_j), j = 1, \ldots, n),$$

where $y_j = f(x_j) + \xi_j$ is the objective function value computed at j iteration, $x_j \in A$, $j = 1, \ldots, n$. It is proven that there is a number $\hat{n}$ so that the inequality $\mathbb{P}(|y_{on} - f_{min}| \leq \delta) > 1 - \epsilon$ is satisfied for $n > \hat{n}$, where $\hat{n}$ depends on f, δ, ϵ and of the variance of ξ_j; see [155]. We do not know about publications on the partition-based Bayesian algorithms for noisy problems with a continuous feasible region.

The performance of an optimization algorithm applied to the true underlying mathematical model of objective functions is especially interesting from the mathematical point of view. Thus, the convergence is analysed as the probabilistic convergence to zero of the error of an approximation of global minimum. The investigation of the probabilistic convergence is indeed challenging, and the results are yet obtained only for univariate algorithms.

The probabilistic convergence of Bayesian algorithms is of interest also for the algorithmic complexity theory [86]. The conclusion of the worst-case analysis of global optimization of Lipschitz functions is pessimistic. Concretely, the optimal sequential algorithm chooses for the worst-case function the same computing points as the optimal passive algorithm [123]. Thus, the optimal algorithms distributes the computation points uniformly in the feasible optimization region. A methodologically interesting question is about the advantage of the sequential planning with respect to the criteria of average efficiency. The analysis of GO algorithms for Wiener process has shown that a sequential algorithms can outperform the optimal passive algorithm. As shown in [102], the uniform search is order-optimal, and its average error Δ_n is of order $n^{-1/2}$. Convergence of Bayesian algorithms in a stochastic setting with respect to the Wiener process model is investigated in a series of papers by Calvin; for a review we refer to [15]. A lower bound of the average error for approximation of the global minimum of paths of Wiener process is established in [13]: there exist such positive c and β that

$$\mathbb{E}(\Delta_n) \geq c \exp(-\beta n / \log(n)), \qquad (2.5.4)$$

is satisfied for any algorithm and all $n \geq 2$. The arbitrarily high polynomial convergence rate can be achieved as shown in [20] by selecting an appropriate parameter of the threshold; more precisely, the following inequality $(\mathbb{E}(\Delta_n^p))^{1/p} \leq cn^{-r}$, can be satisfied for any $1 \leq r < \infty$, by the selection of a corresponding $c > 0$, and the mentioned parameter.

The known bounds for Bayesian algorithms are, although worse than (2.5.4), but reasonably better than the optimal bound for the passive algorithms $\mathbb{E}(\Delta_n) = \Theta(n^{-1/2})$. The convergence in probability of a version of the P-algorithm with a special form of the improvement threshold is proved in [14], and the following estimate of the convergence rate is provided $\mathbb{P}\left(\Delta_n \leq (cn)^{1/4} \exp(-\frac{1}{2}\sqrt{cn})\right) \to 1$, where Δ_n is the error of estimating global minimum after n iterations and $c > 0$ depends on the rate of convergence of the improvement threshold to 0.

For the analysis of the average efficiency of various numerical methods with respect to Wiener process, we refer to [85]. The generalization of these results to multivariate case is indeed challenging, and the general average-case tractability of multivariate global optimization is mentioned as Open Problem 87 in [86].

2.5.3 Hybrid Algorithms

The creation of a unified probabilistic model of objective functions which would well represent desirable global and local properties of aimed objective functions is difficult. Hybridization of different models can be seen as an alternative to the development of an appropriate unified global/local model. The hybridization of algorithms is commonly used to mitigate shortcomings of algorithms having other favourable properties. Bayesian GO algorithms are based on probabilistic models of multimodal objective functions. Local optimization algorithms are designed assuming the unimodality/convexity of objective functions. A potential opportunity arises for testing of hypotheses about adequacy of global or local model over subregions of the feasible region. For example, the dual model combining Wiener process and unimodal function models is used in [149] to hybridize MEI with the local search. During the execution of the MEI algorithm, the statistical hypothesis is tested if a subinterval of unimodality of the objective function is discovered. Such subintervals are excluded from global search, and local minimization is performed over these subintervals. The testing results in [125, 150] have shown the efficiency of that hybrid algorithm in computing the global minimum with the predefined accuracy. However, the generalization of this idea to the multidimensional case seems challenging, whereas it is unclear how to formulate the statistical hypothesis about the unimodality of the sample functions of the underlying GRF over a subset of the feasible region. Nevertheless, a method of approximate description of a probable subregion of the convexity of an objective function, indicated during global search, is proposed in [76].

Methods of computational intelligence can be used for indicating the subregions of the potential global minimizers similarly as the methods of mathematical statistics. Rectangular partition-based Bayesian global search is hybridized with a local minimization algorithm in [162]. The rectangular partition-based P-algorithm [22] is quite advantageous, because of high asymptotic convergence rate and of favourable testing results. On the other hand, after a number of iterations, small hyper-rectangles are further subdivided. Thus, 2^{d-1} function values are computed at closely located points. Very high density of computations in a large part of a basin of potential global minimum is not rational. If a basin of local minimum is discovered, the corresponding minimizer can be refined more efficiently by a local minimization algorithm than continuing global search in that basin. A clustering method is applied in [22] for the indication of discovered subregions of the potential global minima. An indicated during global search potential global minimum is refined by the local descent algorithm. Some neighbourhood of the localized minimizer is excluded from the further global search. The ratio of budgets for exploration and exploitation is more rational for the hybrid algorithm than for its global search counterpart. The experimental results have shown the better performance of the hybrid algorithm than the performance of its global counterpart and corroborated rationality of the hybridization.

The partition-based Bayesian algorithms receive still little attention despite their advantages. Their computational intensity is considerably lower than the computational intensity of the standard implemented Bayesian GO algorithms. The other their advantage is high convergence rate. In addition, the structure of the partition-based algorithms is well suited to the hybridization with local search algorithms.

References

1. S. Agrawal, N. Goyal, Analysis of Thompson sampling for the Multi-armed Bandit problem, in *Proceedings of 25 Conference on Learning Theory*, pp. 39.1—39.26 (2012)
2. A. Aprem, A Bayesian optimization approach to compute the Nash equilibria of potential games using bandit feedback (2018). arXiv:1811.06503v1
3. F. Archetti, A. Candelieri, *Bayesian Optimization and Data Science* (Springer, 2019)
4. P. Auer, Using confidence bounds for exploitation-exploration trade-offs. J. Mach. Learn. Res. **3**, 586–594 (2002)
5. F. Bachoc, C. Helbert, V. Picheny, Gaussian process optimization with failures: classification and convergence proof (2020) HAL Id: hal-02100819
6. R. Bardenet, B. Kegl, Surrogating the surrogate: accelerating Gaussian-process-based global optimization with a mixture cross-entropy algorithm, in *Proceedings of 26 International Conference on Learning Theory*, pp. 1–8 (2010)
7. A. Basudhar, C. Dribusch, S. Lacaze, S. Missoum, Constrained efficient global optimization with support vector machines. Struct. Multidiscip. Optim. **46**, 201–221 (2012)
8. J. Berk, V. Sunil, G. Santu, R. Venkatesh, Exploration enhanced expected improvement for Bayesian optimization. in *Joint European Conference on Machine Learning and Knowledge Discovery in Databases*, pp. 621–637 (2018)
9. H. Bijl, T. Schon, J.-W. van Wingerden, M. Verhaegen, A sequential Monte Carlo approach to Thompson sampling for Bayesian optimization (2017). arXiv:1604.00169v3
10. E. Brochu, V. Cora, and N. de Freitas, A tutorial on Bayesian optimization of expensive cost functions, with application to active user modeling and hierarchical reinforcement learning (2010). arXiv:1012.2599v1
11. A. Bull, Convergence rates of efficient global optimization algorithms. J. Mach. Learn. Res. **12**, 2879–2904 (2011)
12. J. Calvin, Consistency of a myopic Bayesian algorithm for one-dimensional global optimization. J. Glob. Optim. **3**, 223–232 (1993)
13. J. Calvin, A lower bound on complexity of optimization on the Wiener space. Theor. Comput. Sci. **383**, 132–139 (2007)
14. J. Calvin, An adaptive univariate global optimization algorithm and its convergence rate under the wiener measure. Informatica **22**(4), 471–488 (2011)
15. J. Calvin, Probability models in global optimization. Informatica **27**(2), 323–334 (2016)
16. J. Calvin, A. Žilinskas, A one-dimensional P-algorithm with convergence rate $O(n^{-3+\delta})$ for smooth functions. JOTA **106**, 297–307 (2000)
17. J. Calvin, A. Žilinskas, On convergence of a P-algorithm based on a statistical model of continuosly differentiable functions functions. J. Glob. Optim. **19**, 229–245 (2001)
18. J. Calvin, A. Žilinskas, A one-dimensional global optimization for observations with noise. Comp. Math. Appl. **50**, 157–169 (2005)

19. J. Calvin, A. Žilinskas, On a global optimization algorithm for bivariate smooth functions. JOTA **163**(2), 528–547 (2014)
20. J.M. Calvin, M. Hefter, A. Herzwurm, Adaptive approximation of the minimum of Brownian motion. J. Complexity **39**, 17–37 (2017)
21. J.M. Calvin, A. Žilinskas, On the convergence of the p-algorithm for one-dimensional global optimization of smooth functions. JOTA **102**, 479–495 (1999)
22. J. Calvin, et al., On convergence rate of a rectangular partition based global optimization algorithm. J. Glob. Optim. **71**, 165–191 (2018)
23. A. Candelieri, Sequential model based optimization of partially defined functions under unknown constraints. J. Glob. Optim. Published online:1–23 (2019)
24. J. Cashore, L. Kumarga, P. Frazier, Multi-step Bayesian optimization for one-dimensional feasibility determination (2016). arXiv:1607.03195
25. L. Chan, G. Hutchison, G. Morris, Bayesian optimization for conformer generation. J. Cheminformatics **11**(32), 1–11 (2020)
26. B. Chen, R. Castro, A. Krause, Joint optimization and variable selection of high-dimensional Gaussian processes, in *29th International Conference on Machine Learning* (Omnipress, 2012), pp. 1379–1386
27. T. Cormen, C. Leiserson, R. Rivest, C. Stein, *Introduction to Algorithms* (MIT Press, 1989)
28. J. Cui, B. Yang, Survey on Bayesian optimization methodology and applications. J. Softw. **29**(10), 3068–3090 (2007)
29. N. Dolatnia, A. Fern, X. Fern, Bayesian optimization with resource constraints and production, in *Proceedings of 26 International Conference on Automated Planning and Scheduling*, pp. 115–123 (AAAI, 2016)
30. K. Dong et al., Scalable log determinants for Gaussian process kernel learning. Adv. Neural Inf. Proces. Syst. **30**, 6327–6337 (2017)
31. Z. Dou, *Bayesian global optimization approach to the oil well placement problem with quantified uncertainties, Dissertation*. Purdue University (2015)
32. D. Eriksson et al., Scaling Gaussian process regression with derivatives. Adv. Neural Inf. Proces. Syst. **31**, 6867–6877 (2018)
33. D. Eriksson et al., Scalable global optimization via local Bayesian optimization. Adv. Neural Inf. Proces. Syst. **32**, 5496–5507 (2019)
34. Z. Feng et al., A multiobjective optimization based framework to balance the global exploration and local exploitation in expensive optimization. J. Glob. Optim. **61**, 677–694 (2015)
35. P. Frazier, W. Powell, S. Dayanik, The knowledge-gradient policy for correlated normal beliefs. INFORMS J. Comput. **21**(4), 599—613 (2009)
36. J. Gardner, M. Kusner, Z. Xu, K. Weinberger, J. Cunningham, Bayesian optimization with inequality constraints, in *Proceedings of the 31st International Conference on Machine Learning*, pp. II–937–II–945 (2014)
37. R. Garnett, H. Osborne, S. Roberts, Bayesian optimization for sensor set selection, in *Proceedings of International Conference on Information Proceedings Sensor Networks*, pp. 209–219 (2010)
38. M. Gelbart, *Constrained Bayesian Optimization and Applicationss*. Doctoral dissertation, Harvard University (2015)
39. M. Gelbart, J. Snoek, R. Adams, Bayesian optimization with unknown constraints, in *Proceedings of 30 conference on Uncertainty in AI*, pp. 250–259 (2014)
40. E. Gilboa, Y. Saatci, J. Cunningham, Scaling multidimensional Gaussian processes using projected additive approximations, in *Proceedings of the 30 International Conference on Machine Learning*, vol. 28 (2013), pp. I–454–I–461
41. D. Ginsbourger, J. Janusevskis, R. Le Riche, Dealing with asynchronicity in parallel Gaussian process based global optimization (2011). HAL Id: hal-00507632
42. D. Ginsbourger, R. Le Riche, Towards GP-based optimization with finite time horizon (2009). https://hal.archives-ouvertes.fr/hal-00424309/en/
43. R. Gramacy, H. Lee, Optimization under unknown constraints. Bayesian Stat. **9**, 1–18 (2011)

44. R.B. Gramacy, J. Niemi, R.M. Weiss, Massively parallel approximate Gaussian process regression. SIAM/ASA J. Uncertain. Quantif. **2**(1), 564–584 (2014)
45. R. Griffiths, Constrained Bayesian Optimization for Automatic Chemical Design. Dissertion, University of Cambridge (2017)
46. R. Griffiths, J. Hernandez-Lobato, Constrained Bayesian optimization for automatic chemical design (2019). arXiv:1709.05501v6
47. Z. Han, M. Abu-Zurayk, S. G̈örtz, C. Ilic, Surrogate-based aerodynamic shape optimization of awing-body transport aircraft configuration, in *Notes on Numerical Fluid Mechanics and Multidisciplinary Design*, vol. 138 (Springer, 2018), pp. 257–282
48. F. Hase et al., Phoenics: A Bayesian optimizer for chemistry. ACS Cent. Sci. **4**, 1134–1145 (2020)
49. P. Hennig, C. Schuler, Entropy search for information-efficient global optimization. J. Mach. Learn. Res. **13**, 1809–1837 (2012)
50. J. Hernandez-Lobato, M. Gelbart, R. Adams, M. Hofman, Z. Ghahramani, A general framework for constrained Bayesian optimization using information-based search. J. Mach. Learn. Res. **17**, 1–53 (2016)
51. J. Hernandez-Lobato, M. Gelbart, M. Hofman, R. Adams, Z. Ghahramani, Predictive entropy search for Bayesian optimization with unknown constraints (2015). arXiv:1502.05312v2
52. J. Hernandez-Lobato, J. Requeima, E. Pyzer-Knapp, A. Aspuru-Guzik, Parallel and distributed Thompson sampling for large-scale accelerated exploration of chemical space (2017). arXiv:1706.01825v1
53. D. Huang, T. Allen, W. Notz, R. Miller, Sequential kriging optimization using multiple-fidelity evaluations. Struct. Multidiscip. Optim. **32**, 369—382 (2006)
54. H. Jalali, I. Nieuwenhuyse, V. Picheny, Comparison of kriging-based algorithms for simulation optimization with heterogeneous noise. EJOR **261**(1), 279–301 (2017)
55. S. Jeong, M. Murayama, K. Yamamoto, Efficient optimization design method using kriging model. J. Aircraft **42**(2), 413–422 (2005)
56. D.R. Jones, C.D. Perttunen, C.D. Stuckman, Lipschitzian optimization without the Lipschitz constant. JOTA **79**, 157–181 (1993)
57. D.R. Jones, M. Schonlau, W. Welch, Efficient global optimization of expensive black-box functions. J. Glob. Optim. **13**, 455–492 (1998)
58. K. Kandasamy, A. Krishnamurthy, J. Schneider, B. Poczos, Parallelised Bayesian optimisation via Thompson sampling, in *Proceedings of 21 International Conference on Artificial Intelligence and Statistics*, pp. 1–10 (2018)
59. J. Kim, S. Choi, Clustering-guided GP-UCB for Bayesian optimization, in *IEEE International Conference on Acoustics, Speech, and Signal Processing*, pp. 2461–2465 (2018)
60. T. Kim, J. Lee, Y. Choe, Bayesian optimization-based global optimal rank selection for compression of convolutional neural networks. IEEE Access **8**, 17605–17618 (2020)
61. J. Kleijnen, W. van Beers, I. van Nieuwenhuyse, Expected improvement in efficient global optimization through bootstrapped kriging. J. Glob. Optim. **54**, 59–73 (2012)
62. J. Knowles, D. Corne, A. Reynolds, Noisy multiobjective optimization on a budget of 250 evaluations, in *Lecture Notes in Computer Science*, ed. by M. Ehrgott et al. vol. 5467 (Springer, 2009), pp. 36–50
63. H. Kushner, A versatile stochastic model of a function of unknown and time-varying form. J. Math. Anal. Appl. **5**, 150–167 (1962)
64. H. Kushner, A new method of locating the maximum point of an arbitrary multipeak curve in the presence of noise. J. Basic Eng. **86**, 97–106 (1964)
65. R. Lam, M. Poloczeky, P. Frazier, K. Willcox, Advances in Bayesian optimization with applications in aerospace engineering, in *AIAA Non-Deterministic Approaches Conference*, pp. 1–10 (2018)
66. R. Lam, K. Willcox, Lookahead Bayesian optimization with inequality constraints, in *31st Conference on Neural Information Processing Systems*, pp. 4–5 (2017)
67. L. Cornejo-Buenoa, E.C. Garrido-Merchánb, D. Hernández-Lobatob, S. Salcedo-Sanza, Bayesian optimization of a hybrid system for robust ocean wave features prediction. Neurocomputing **275**, 818–828 (2018)

68. B. Letham, B. Karrery, G. Ottoniz, E. Bakshy, Constrained Bayesian optimization with noisy experiments (2018) arXiv:1706.07094v2
69. C. Li, S. Gupta, S. Rana, V. Nguyen, S. Venkatesh, A. Shilton, High dimensional Bayesian optimization using dropout, in *Proceedings of 26 International Conference on AI*, pp. 2096–2102 (2017)
70. C. Li, K. Kandasamy, B. Poczos, J. Schneider, High dimensional Bayesian optimization via restricted projection pursuit models, in *Proceedings of 19 International Conference on Artificial Intelligence and Statistics* (Springer, 2016), pp. 884–892
71. D. Lindberg, H.K. Lee, Optimization under constraints by applying an asymmetric entropy measure. J. Comput. Graph. Stat. **24**(2), 379–393 (2015)
72. W.-L. Loh, T.-K. Lam, Estimating structured correlation matrices in smooth Gaussian random field models. Ann. Stat. **28**, 880–904 (2000)
73. M. Maier, A. Rupenyan1, C. Bobst, K. Wegener, Self-optimizing grinding machines using Gaussian process models and constrained Bayesian optimization (2020). arXiv:2006.05360v1
74. A. Makauskas, On a possibility to use gradients in statistical models of global optimization of objective functions. Informatica **2**, 248–254 (1991)
75. G. Malkomes, R. Garnett, Automating Bayesian optimization with Bayesian optimization, in *32 Conference on Neural Information Processing Systems*, pp. 1–11 (2018)
76. M. McLeod, M. Osborne, S. Roberts, Optimization, fast and slow: optimally switching between local and Bayesian optimization (2018). arXiv:1805.08610v1
77. A. Mittal, S. Aggarwal, Hyperparameter optimization using sustainable proof of work in blockchain. Front. Blockchain **3**(23), 1–13 (2020)
78. J. Mockus, On Bayes methods for seeking an extremum. Avtomatika i Vychislitelnaja Technika (3), 53–62 (1972) in Russian
79. J. Mockus, *Bayesian Approach to Global Optimization* (Kluwer Academic Publishers, 1988)
80. J. Mockus, V. Tiešis, A. Žilinskas, The application of Bayesian methods for seeking the extremum, in *Towards Global Optimization 2*, ed. by L.C.W. Dixon, G.P. Szego (North Holland, 1978), pp. 117–129
81. J. Mockus et al., *Bayesian Heuristic Approach to Discrete and Global Optimization* (Kluwer Academic Publishers, Dodrecht, 1997)
82. M. Morrar, J. Knowles, S. Sampaio, Initialization of Bayesian optimization viewed as part of a larger algorithm portfolio, in *CEC2017 and CPAIOR 2017*, pp. 1–6 (2017)
83. M. Mutny, A. Krause, Efficient high dimensional Bayesian optimization with additivity and quadrature Fourier features, in *32 Conference on Neural Information Processing Systems*, pp. 1–12 (2018)
84. V. Nguyen et al., Regret for expected improvement over the best-observed value and stopping condition, in *Proceedings of 9 Asian Conference on Machine Learning*, vol. 77 (PMLR, 2017), pp. 279–294
85. E. Novak, *Deterministic and Stochastic Error Bounds in Numerical Analysis*, volume 1349 of *Lecture Notes in Mathematics* (Springer, Berlin, 1988)
86. E. Novak, H. Woźniakowski, *Tractability of Multivariate Problems*, volume II of *Tracts in Mathematics* (European Mathematical Society, Zürich, 2010)
87. S. Olofsson et al., Bayesian multiobjective optimisation with mixed analytical and black-box functions: Application to tissue engineering. IEEE Trans. Biomed. Eng. **66**(3), 727–739 (2019)
88. M. Osborne, R. Garnett, S. Roberts, Gaussian processes for global optimization (2009). http://www.robots.ox.ac.uk
89. Y. Ozaki et al., Automated crystal structure analysis based on blackbox optimisation. Comput. Mat. **6**(75), 1–7 (2020)
90. R. Paulavičius et al., Globally-biased Disimpl algorithm for expensive global optimization. J. Glob. Optim. **59**, 545–567 (2014)
91. V. Picheny, Multiobjective optimization using gaussian process emulators via stepwise uncertainty reduction. Stat. Comput. **25**, 1265–1280 (2015)

92. V. Picheny, D. Ginsbourger, Y. Richet, Noisy expected improvement and on-line computation time allocation for the optimization of simulators with tunable fidelity, in *Proceedings of 2nd International Conference on Engineering Opt* (2010)

93. V. Picheny, R. Gramacy, S. Wild, S. Le Digabel, Bayesian optimization under mixed constraints with a slack-variable augmented lagrangian (2016). arXiv:1605.09466v1

94. V. Picheny, T. Wagner, D. Ginsbourger, A benchmark of kriging-based infill criteria for noisy optimization. Struct. Multidiscip. Optim. **48**(3), 607—626 (2013)

95. J. Pinter, *Global Optimization in Action* (Kluwer Academic Publisher, 1996)

96. R. Preuss, U. von Toussaint, Global optimization employing Gaussian process-based Bayesian surrogates. Entropy **20**, 201–214 (2018)

97. R. Priem et al., An adaptive feasibility approach for constrained Bayesian optimization with application in aircraft design, in *6 International Conference on Engineering Optimization (EngOpt2018)* (2018)

98. H. Prosper, Deep learning and Bayesian methods. EPJ Web Conf. **137**, 11007 (2018)

99. S. Rana, C. Li, S. Gupta, V. Nguyen, S. Venkatesh, High dimensional Bayesian optimization with elastic Gaussian process, in *Proceedings of 34th International Conference on Machine Learning*, pp. 2883–2891 (2017)

100. C.E. Rasmussen, C. Williams, *Gaussian Processes for Machine Learning* (MIT Press, 2006)

101. B. Rezaeianjouybari, M. Sheikholeslami, A. Shafee, H. Babazadeh, A novel Bayesian optimization for flow condensation enhancement using nanorefrigerant: A combined analytical and experimental study. Chem. Eng. Sci. **215**, 115465 (2020)

102. K. Ritter, Approximation and optimization on the Wiener space. J. Complexity **6**, 337—364 (1990)

103. M. Sacher et al., A classification approach to efficient global optimization in presence of non-computable domains. Struct. Multidiscip. Optim. **58**(4), 1537–1557 (2018)

104. J. Sacks, S.B. Schiller, W.J. Welch, Designs for computer experiments. Technometrics **31**(1), 41–47 (1989)

105. J. Sacks, W.J. Welch, T.J. Mitchell, H.P. Wynn, Design and analysis of computer experiments. Stat. Sci. **4**, 409–423 (1989)

106. M. Sasena, *Dissertation: Flexibility and Efficiency Enhancements for Constrained Global Design Optimization with Kriging Approximations*. Michigan University (2002)

107. M. Schonlau, W. Welch, D. Jones, *Global versus local search in constrained optimization of computer models, Technical Report Number 83*. National Institute of Statistical Sciences (1998)

108. Y. Sergeyev, An efficient strategy for adaptive partition of n-dimensional intervals in the framework of diagonal algorithms. JOTA **107**, 145–168 (2000)

109. Y. Sergeyev, D. Kvasov, Global search based on efficient diagonal partitions and a set of Lipschitz constants. SIAM J. Optim. **16**, 910–937 (2006)

110. Y. Sergeyev, Numerical infinities and infinitesimals: Methodology, applications, and repercussions on two Hilbert problems. EMS Surv. Math. Sci. **4**, 219–320 (2017)

111. B. Shahriari, K. Swersky, Z. Wang, R. Adams, N. de Freitas, Taking the human out of the loop: A review of Bayesian optimization. Proc. IEEE **104**(1), 148–175 (2016)

112. B. Shahriari, Z. Wang, M. Hoffman, An entropy search portfolio for Bayesian optimization (2015). arXiv:1406.4625v4

113. V. Shaltenis, On a method of multiextremal optimization. Avtomatika i Vychislitelnaja Technika **3**, 33–38 in Russian (1971)

114. D. Silveret et al., Mastering the game of Go with deep neural networks and tree search. Nature **529**, 484—489 (2016)

115. T. Simpson, J. Korte, F. Mistree, Kriging models for global approximation in simulation-based multidisciplinary design optimization. AIAA J. **39**(12), 2233–2242 (2001)

116. N. Srinivas, A. Krause, S. Kakade, M. Seeger, Gaussian process optimization in the Bandit setting: No regret and experimental design, in *Proceedings of 27th International Conference on Machine Learning*, pp. 1015–1022 (Omnipress, 2010)

117. M.L. Stein, *Interpolation of Spatial Data: Some Theory of Kriging* (Springer, 1999)

118. L. Stripinis, R. Paulavičius, J. Žilinskas, Improved scheme for selection of potentially optimal hyper-rectangles in DIRECT. Optim. Lett. **12**, 1699–1712 (2018)
119. R. Strongin, Information method of global minimization in the presence of noise. Eng. Cybern. **6**, 118–126 (1969) in Russian
120. R.G. Strongin, *Numerical Methods of Multiextremal Minimization*. Nauka, (1978) in Russian
121. R.G. Strongin, Y.D. Sergeyev, *Global Optimization with Non-convex Constraints: Sequential and Parallel Algorithms* (Kluwer Academic Publishers, 2000)
122. A. Sukharev, *Minimax Models in the Theory of Numerical Methods* (Springer, 2012)
123. A.G. Sukharev, Optimal strategies of search for an extremum. USSR Comput. Math. Math. Phys. **11**(4), 910–924 (1971)
124. R. Tamura, K. Hukushima, Bayesian optimization for computationally extensive probability distributions. PLoS ONE **13**, e0193785 (2018)
125. A. Törn, A. Žilinskas, *Global Optimization* (Springer, 1989)
126. E. Vazquez, J. Bect, Convergence properties of the expected improvement algorithm with fixed mean and covariance functions. J. Stat. Plan. Infer. **140**(11), 3088–3095 (2010)
127. N. Vien, H. Zimmermann, M. Toussaint, Bayesian functional optimization, in *32 AAAI Conference on AI*, pp. 4171–4178 (AAAI, 2018)
128. J. Villemonteix, E. Vazquez, E. Walter, An informational approach to the global optimization of expensive to evaluate functions. J. Glob Optim. **44**(4), 509–534 (2009)
129. K. Wabersich, Gaussian processes and Bayesian optimization (2016)
130. K. Wabersich, M. Toussaint, Advancing Bayesian optimization: The mixed-global-local kernel and length-scale cool down (2016). arXiv:1612.03117v1
131. J. Wang, S. Clark, E. Liu, P. Frazier, Parallel Bayesian global optimization of expensive functions (2019). arXiv:1602.05149v4
132. Z. Wang, S. Jagelka, Max-value entropy search for efficient Bayesian optimization (2018). arXiv:1703.01968v3
133. K. Wang et al., Exact Gaussian processes on a million data points, in *33rd Conference on Neural Information Processing Systems*, pp. 1—13 (2019). arXiv:1903.08114v2
134. Z. Wang et al., Bayesian optimization in a billion dimensions via random embeddings. J. AI Res. **55**, 361–387 (2016)
135. Z. Wang et al., Bayesian optimization in high dimensions via random embeddings, in *Proceedings of 23 International Conference on AI*, pp. 1778–1784 (2017)
136. J. Wilson, V. Borovitskiy, A. Terenin, P. Mostowsky, M. Deisenroth, Efficiently sampling functions from gaussian process posteriors (2020). arXiv:2002.09309v1
137. J. Wilson, F. Hutter, M. Deisenroth, Maximizing acquisition functions for Bayesian optimization (2018). arXiv:1805.10196v2
138. A. Wu, M. Aoi, J. Pillow, Exploiting gradients and Hessians in Bayesian optimization and Bayesian quadrature (2018). arXiv:1704.00060v2
139. J. Wu, P. Frazier, Discretization-free Knowledge Gradient methods for Bayesian optimization (2017). arXiv:1707.06541v1
140. J. Wu, M. Poloczek, A. Wilson, P. Frazier, Bayesian optimization with gradients, in *Proceedings of 31st International Conference on Neural Information Processing Systems*, pp. 5273–5284 (IEEE, 2017)
141. J. Wu et al., Hyperparameter optimization for machine learning models based on Bayesian optimization. J. Electron. Sci. Technol. **17**(1), 26–40 (2019)
142. W. Xu, M.L. Stein, Maximum likelihood estimation for smooth Gaussian random field model. SIAM/ASA Uncertain. Quantif. **5**, 138–175 (2017)
143. A.M. Yaglom, *Correlation Theory of Stationary and Related Random Functions,* vol. 1 (Springer, 1987)
144. J. Yim, B. Lee, C. Kim, Exploring multi-stage shape optimization strategy of multi-body geometries using kriging-based model and adjoint method. Comput. Fluids **68**, 71–87 (2012)
145. A. Zhigljavsky, A. Žilinskas. *Methods of Search for Global Extremum* (Nauka, Moscow, 1991), in Russian
146. A. Zhigljavsky, A. Žilinskas, *Stochastic Global Optimization* (Springer, 2008)

147. A. Zhigljavsky, A. Žilinskas, Selection of a covariance function for a Gaussian random field aimed for modeling global optimization problems. Opt. Lett. **13**(2), 249—259 (2019)

148. A. Žilinskas, One-step Bayesian method for the search of the optimum of one-variable functions. Cybernetics 1, 139–144 (1975) in Russian

149. A. Žilinskas, On global one-dimensional optimization. Izv. Acad. Nauk USSR Eng. Cybern. **4**, 71–74 (1976) in Russian

150. A. Žilinskas, Optimization of one-dimensional multimodal functions, algorithm 133. J. Roy. Stat. Soc. Ser C **23**, 367–385 (1978)

151. A. Žilinskas, MIMUN-optimization of one-dimensional multimodal functions in the presence of noise. Aplikace Matematiky **25**, 392–402 (1980)

152. A. Žilinskas, Two algorithms for one-dimensional multimodal minimization. Math. Oper. Stat. Ser. Optim. **12**, 53–63 (1981)

153. A. Žilinskas, Axiomatic approach to statistical models and their use in multimodal optimization theory. Math. Program. **22**, 104–116 (1982)

154. A. Žilinskas, Axiomatic characterization of a global optimization algorithm and investigation of its search strategies. Oper. Res. Lett. **4**, 35–39 (1985)

155. A. Žilinskas, *Global Optimization: Axiomatic of Statistical Models, Algorithms, Applications* (Mokslas, Vilnius, 1986) in Russian

156. A. Žilinskas, Statistical models for global optimization by means of select and clone. Optimization **48**, 117–135 (2000)

157. A. Žilinskas, On the worst-case optimal multi-objective global optimization. Optim. Lett. **7**(8), 1921–1928 (2013)

158. A. Žilinskas, Including the derivative information into statistical models used in global optimization. AIP Conf. Proc. **2070**(020020), 1–4 (2019)

159. A. Žilinskas, A. Makauskas, On possibility of use of derivatives in statistical models of multimodal functions, in *Teorija Optimaljnych Reshenij*, vol. 14, pp. 63–77. Inst. Math. Cybern. Lith. Acad. Sci. (1990) in Russian

160. A. Žilinskas, J.M. Calvin, Bi-objective decision making in global optimization based on statistical models. J. Glob. Optim. **74**, 599–609 (2019)

161. A. Žilinskas, G. Gimbutienė, On asymptotic property of a simplicial statistical model of global optimization, in *Springer Proceedings in Mathematics and Statistics*, vol. 130 (2015), pp. 383–392

162. A. Žilinskas, G. Gimbutienė, A hybrid of Bayesian approach based global search with clustering aided local refinement. Commun. Nonlinear Sci. Numer. Simul. **78**, 104857 (2019)

163. A. Žilinskas, L. Litvinas, A hybrid of the simplicial partition-based bayesian global search with the local descent. Soft Comput. **24**, 17601–17608 (2020)

164. A. Žilinskas, J. Mockus, On a Bayesian method for seeking the minimum. Avtomatika i Vychislitelnaja Technika **4**, 42–44 (1972) in Russian

165. A. Žilinskas, E. Senkiene, On estimating the parameter of Wiener process. Lith. Math. J. **3**, 59–62 (1978) in Russian

166. A. Žilinskas, J. Žilinskas, Global optimization based on a statistical model and simplicial partitioning. Comput. Math. Appl. **44**(7), 957–967 (2002)

167. A. Žilinskas, J. Žilinskas, P-algorithm based on a simplicial statistical model of multimodal functions. TOP **18**, 396–412 (2010)

168. A. Žilinskas et al., Multi-objective optimization and decision visualization of batch stirred tank reactor based on spherical catalyst particles. Nonlinear Anal. Model. Control **24**(6), 1019–1036 (2019)

Chapter 3
Global Random Search in High Dimensions

It is not the aim of this chapter to cover the whole subject of the global random search (GRS). It only contains some potentially important notes on algorithms of GRS in continuous problems, mostly keeping in mind the use of such algorithms in reasonably large dimensions. These notes are based on the 40-year experience of the author of this chapter and reflect his subjective opinions, some of which other specialists in the field do not necessarily share. The chapter discusses new results as well as rather old ones thoroughly explained in the monographs [49, 51, 53].

3.1 Main Ideas, Principles and Classes of Algorithms

A generic GRS algorithm assumes that a sequence of random points $x_1, x_2, \ldots, x_n$ is generated where for each $j \geqslant 1$ the point x_j has some probability distribution P_j (we write this $x_j \sim P_j$). For each $j \geqslant 2$, the distribution P_j may depend on the previous points $x_1, \ldots, x_{j-1}$ and on the results of the objective function evaluations at these points (the function evaluations may not be noise-free). The number of points n (the stopping rule) can be either deterministic or random and may depend on the results of function evaluation at the points $x_1, \ldots, x_n$; see Sect. 3.3.1. In order for an algorithm to be classified as a GRS algorithm, at least one of the distributions P_j should be non-degenerate (i.e. at least one of x_j is a random point in $\mathscr{X}$). In contrast to the Bayesian methods of Chap. 2, the underlying assumption behind GRS algorithms is the assumption that evaluations of the objective function $f(\cdot)$ are not expensive. GRS algorithms are based on the black-box assumption about $f(\cdot)$; any knowledge of the structure of the optimization problem can be used for construction of effective optimization algorithms; for examples, see some references at the end of Sect. 3.4.

© The Author(s) 2021

A. Zhigljavsky, A. Žilinskas, *Bayesian and High-Dimensional Global Optimization*,
SpringerBriefs in Optimization, https://doi.org/10.1007/978-3-030-64712-4_3

Many GRS algorithms have the following attractive features: (a) the structure of GRS algorithms is usually simple; (b) these algorithms are often rather insensitive to the irregularity of the objective function behaviour, to the shape of the feasible region, to the presence of noise in the objective function evaluations and even to the growth of dimensionality; (c) it is very easy to construct GRS algorithms guaranteeing theoretical convergence.

The drawbacks of GRS algorithms are (a) practical efficiency of GRS algorithms often depends on a number of parameters, but the problem of the choice of these parameters frequently has little relevance to the theoretical results concerning the convergence of the algorithms; (b) for many GRS algorithms, an analysis on good parameter values is lacking or simply impossible; (c) the theoretical convergence rate is extremely slow; see Sect. 3.1.3.

The theoretical convergence rate is expressed in terms of the covering properties of the sequence of points generated by an algorithm; such properties are studied in Sect. 1.3. In the class of genuine GRS algorithms, the simplest algorithms like the PRS defined below on page 91 possess the best possible convergence rates and are not improvable. On the other hand, improving efficiency of algorithms in specific classes of optimization problems is perfectly possible, and this is the main subject matter of the theory and methodology of GRS.

3.1.1 Main Principles and Important Classes of GRS Algorithms

A very large number of specific GRS algorithms exist, but only a few main principles form their basis. These principles can be summarized as follows:

P1: random sampling of points at which $f(\cdot)$ is evaluated,
P2: covering of the space (exploration),
P3: combination with local optimization techniques (exploitation),
P4: use of different heuristics including cluster-analysis techniques to avoid clumping of points around particular local minimizers,
P5: use of statistical inference,
P6: more frequent selection of new trial points in the vicinity of 'good' previous points, and
P7: decrease of randomness in the selection rules for the trial points.

Principle P1 classifies an optimization algorithm as a GRS algorithm. P2 makes sure that that the search is global, while P3 looks after local improvements in the process of search. Good local solutions improve the record values, which work as thresholds for the new points and are used in many algorithms for defining the prospectiveness of subsets of $\mathscr{X}$ for further search.

A right balance between globality (exploration) and locality (exploitation) of search is one of the main ingredients of algorithm's efficiency. Achieving the right balance depends on complexity of computing derivatives of $f(\cdot)$ (for performing fast local descent) and on efficient use of all available information (prior information and information obtained during the process of search) about $f(\cdot)$ and $\mathscr{X}$. Processing the information about $f(\cdot)$ and $\mathscr{X}$, obtained during the process of search, can be achieved by the methodologies associated with Principles P4, P5 and P6. The standard reference for P4 is [38]; see also Sect. 3.1.2. Principles P5 and P6 are subject matters of Sects. 3.2 and 3.3, respectively. Principle P7 is discussed in Sect. 3.2.4. It is argued in that section that any suitable decrease of randomness in the choice of points x_j leads to better (i.e. more efficient) optimization algorithms.

Below we distinguish five popular classes of GRS algorithms with simple updating rules for the distributions P_j.

PRS (*Pure Random Search*). Random points $x_1, x_2, \ldots$ are independent and have the same distribution: $x_j \sim P$ (so that $P_j = P$ for all j); see Sect. 3.1.3.

MGS (*Markovian Global Search*). The distribution P_j depends on x_{j-1} and the objective function value at this point but does not depend on the values of $f(\cdot)$ computed earlier.

PAS (*Pure Adaptive Search*). P_j is uniform on the set $\mathscr{X}_j = \{x \in \mathscr{X} : f(x) \leq y_{0,j-1}\}$, where $y_{0,j-1} = \min_{i=1\ldots j-1} f(x_i)$ is the record value at time $j-1$.

PBS (*Population-Based Search*) Similar to MGS but groups (populations) of points are probabilistically transformed into subsequent groups rather than points to points in MGS, Sect. 3.3.

RMS (*Random Multi-Start*). Local descents are performed from a number of random points in $\mathscr{X}$; see Sect. 3.1.2.

Simplicity of PRS allows detailed investigation of this algorithm; see Sects. 3.1.3 and 3.2. Note, however, the results of Sect. 1.3.4 imply that *in high dimensions the choice of the uniform distribution for generating points in $\mathscr{X}$ is far from being optimal, in contrast to the common belief.*

MGS algorithms, including the simulated annealing [47], are more clever than the primitive PRS. At the same time, MGS algorithms are simple enough to allow a thorough theoretical investigation. There are many papers on simulated annealing and other MGS, but the practical efficiency of these algorithms is rather poor: indeed, MGS algorithms are too myopic, and they waste almost all information about the objective function which is collected in the process of search; see [49, 51] for more detailed arguments. For a more positive view on the convergence rate of MGS algorithms, see [42–44]. MCS algorithms have much resemblance with the MCMC algorithms widely used in Bayesian statistics; see, e.g. [39]. MGS algorithms can be naturally generalized so that the distributions P_j depend not only on x_{j-1} and $f(x_{j-1})$ but also on the current record $y_{0,j-1}$ and the point where this record has been computed. Practically, these algorithms can be much more efficient than the pure MGS algorithm, but the theoretical study of these algorithms is generally very complicated in view of dependence of P_j on all previous points. An obvious exception is PAS considered next.

The idea of PAS is extremely simple, and different versions of PAS have been known long before publication of [30], where PAS has received its name and high (but short-lived) popularity; see, for example, the book [48] almost exclusively devoted to PAS. Popularity of PAS is due to the fact that under some natural assumptions about $f(\cdot)$ and $\mathscr{X}$, PAS converges linearly. However, generation of random points in the sets $\mathscr{X}_j = \{x \in \mathscr{X} : f(x) \leqq y_{0,j-1}\}$ is extremely difficult. In particular, if to obtain a uniform random point in $\mathscr{X}_j$ one generates independent random points in $\mathscr{X}$ and waits for the arrival of a point at $\mathscr{X}_j$, then, as will be made clear in Sect. 3.2, the expected waiting time is infinite, even for $j = 2$! The trouble for PAS is that there is no other general way of getting random points in $\mathscr{X}_j$. Moreover, as shown in [21], if PAS would be realizable in a time that grows polynomially in the dimension, then we would be able to solve an NP-hard problem in an expected number of iterations which grows polynomially in the dimension of the problem.

In PBS methods, populations (i.e. bunches of points) evolve rather than individual points. There is a lot of literature on PBS, but the majority of publications devoted to PBS deal with meta-heuristics rather than with theory and generic methodology. Several probabilistic models where populations are associated with probability distributions are proposed and investigated in Chap. 5 of [51]. We will briefly consider some theoretical issues related to PBS in Sect. 3.3.

RMS is an extremely popular algorithm in practical optimization. It is very clear and very easy to implement. Typical multistart descent iterations initialized at random starting points are displayed in Fig. 3.1. Efficiency of RMS depends on the efficiency of the local search routines, the number of its local minimizers of $f(\cdot)$ and the volumes of different regions of attraction of local minimizers. Repeated descents to the same local minimizers is the obvious disadvantage of multistart. It is usually advisable to use the clustering heuristic of [38] in order to avoid clumping of points around local minimizers; for later results and references, see [46]. In Sect. 3.1.2 we discuss some statistical procedures that could substantiate the use of RMS and outline a version multistart, where most of the descents are terminated very early, much earlier than reaching a close neighbourhood of a local minimizer.

Note that multistart can easily be made more efficient if points with better covering properties (see Sect. 1.3) are used as the starting points for local descents. For these versions of multistart, however, we may not be able to use the statistical procedures discussed below in Sects. 3.1.2, 3.1.3, and 3.2.

Certain classes of practical problems create global optimizations problems in super-high dimensions with the additional difficulty that the objective function is evaluated with an error. A very notable class of such optimization problems is the construction of deep learning networks. In the case of super-high dimensions, it is hard to suggest something more clever than a version of the stochastic gradient algorithm; see, e.g. [24]. Note, however, that the stochastic gradient in high dimensions can often be improved using the approach developed in [17, 18]. The approach of [17, 18] recommends the use of simple stochastic directions $X^\top Y$ rather than least-square estimators $(X^\top X)^{-1} X^\top Y$ of the gradient of the objective function, where we use the standard notation of linear regression analysis with X being the design matrix at chosen coordinates; these coordinates change at each iteration. The directions $X^\top Y$ are much simpler than the estimators of the gradient and maximize the probability that the chosen direction improves the objective function value.

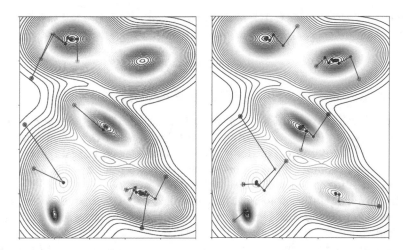

Fig. 3.1 Typical steepest decent iterations

3.1.2 Multistart

Multistart with early termination of descents In this discussion, we introduce the algorithm developed in [56]. The main idea of the algorithm is based on the following result on monotonicity of descent trajectories for quadratic objective functions.

Assume that we have a quadratic function

$$f(x) = \frac{1}{2} x^T A x + b^T x + c, \quad x \in \mathbb{R}^d, \tag{3.1.1}$$

where A is a positive definite $d \times d$ matrix, b is some vector in $\mathbb{R}^d$ and c is some constant. The gradient of $f(\cdot)$ at x is $\nabla f(x) = Ax + b$. In this case, given a point $x_k \in \mathbb{R}^d$, a k-th iteration of a gradient descent algorithm would return the point

$$x_{k+1} = x_k - \gamma_k \nabla f(x_k) = x_k - \gamma_k (Ax_k + b), \tag{3.1.2}$$

where $\gamma_k \geq 0$ is some step-size.

Theorem 3.1 (see [56]) *Let $f(\cdot)$ be a quadratic function (3.1.1), where A is a positive definite matrix. Let x and y be two arbitrary points in $\mathbb{R}^d$ such that $\|\nabla f(x)\| > 0$ and $\|\nabla f(y)\| > 0$. Fix some $\beta > 0$ and define*

$$\begin{cases} \tilde{x} = x - \beta \nabla f(x), \\ \tilde{y} = y - \beta \nabla f(y); \end{cases} \tag{3.1.3}$$

that is, we apply the rule (3.1.2) to the points x and y with the same step-size β. If
β < 1/λ_max, where λ_max is the maximal eigenvalue of the matrix A, then

$$\|\tilde{x} - \tilde{y}\| < \|x - y\|, \tag{3.1.4}$$

where $\|\cdot\|$ denotes the Euclidean norm of a vector in $\mathbb{R}^d$.

We will call the points $\tilde{x}$ and $\tilde{y}$ computed by the rule (3.1.3) 'the partner points associated with x and y respectively'. The theorem above can be interpreted as saying that if the objective function f is quadratic and the coefficient β in (3.1.3) is small enough then for arbitrary x and y, the associated partner points $\tilde{x}$ and $\tilde{y}$ are always closer to each other than the original points x and y.

This is totally different to what happens when the partner points $\tilde{x}$ and $\tilde{y}$ are computed for gradients of two different functions. Indeed, assume we have two quadratic functions $f_i(x) = \frac{1}{2}x^T A_i x + b_i^T x + c_i$ $(i = 1, 2)$, $x \in \mathbb{R}^d$, where A_1 and A_2 are two different non-negative definite $d \times d$ matrices, b_1 and b_2 are two vectors in $\mathbb{R}^d$ and c_1 and c_2 are some constants.

For two arbitrary points, x and y in $\mathbb{R}^d$ define their partner points by

$$\tilde{x} = x - \beta \nabla f_1(x) = x - \beta(A_1 x + b_1),$$
$$\tilde{y} = y - \beta \nabla f_2(y) = y - \beta(A_2 y + b_2).$$

Then $\tilde{x} - \tilde{y} = (x - y) - \beta(A_1 x + b_1 - A_2 y - b_2)$. If we impose some natural randomness assumptions on either points x and y, vectors b_1 and b_2 or matrices A_1 and A_2, then we may observe that the inequality (3.1.4) holds with probability much smaller than 1. In fact, in the simulation studies we have performed, this probability was rather close to 0, especially for large dimensions.

Theorem 3.1 and the assumption of local quadraticity of the objective function is the base of the algorithm developed in [56], which has proved to work very well at the global optimization problems with a black-box objective function $f(\cdot)$ characterized by the following properties:

(a) the feasible domain $\mathfrak{X}$ is high-dimensional but has simple structure;
(b) $\|\nabla f(x)\| \neq 0$ for almost all $x \in \mathfrak{X}$, where $\nabla f(x)$ is the gradient of $f(\cdot)$;
(c) computation of the objective function values and its derivatives is not expensive;
(d) the total number of local minimizers is not very large;
(e) the volume of the region of attraction of the global minimizer is not very small.

The algorithm has been named METOD as an abbreviation for 'multistart with early termination of descents'.

Statistical inferences in RMS In this discussion, we recall some results of R. Zieliński published in a seminal paper [54], which can be used to devise clever stopping rules in random multistart. Note that there has been very little progress in the area of making statistical inferences in random multistart since 1981, the time of the publication of [54].

Assume that vol($\mathscr{X}$)=1, $f(\cdot)$ has a finite but unknown number ℓ of local minimizers $x_*^{(1)}, \ldots, x_*^{(\ell)}$, and $\mathscr{A}$ be a local descent algorithm. We write $\mathscr{A}(x) = x_*^{(i)}$ for $x \in \mathscr{X}$, if starting at the initial point x the algorithm $\mathscr{A}$ leads to the local minimizer $x_*^{(i)}$. Set $\theta_i = \text{vol}(A_i)$ for $i = 1, \ldots, \ell$, where $A_i = \{x \in \mathscr{X} : \mathscr{A}(x) = x_*^{(i)}\}$ is the region of attraction of $x_*^{(i)}$. It follows from the definition that $\theta_i > 0$ for $i = 1, \ldots, \ell$ and $\sum_{i=1}^{\ell} \theta_i = 1$.

The simplest version of RMS is the following simple algorithm: an independent sample $\mathbb{Z}_N = \{x_1, \ldots, x_N\}$ from the uniform distribution on $\mathscr{X}$ is generated and a local optimization algorithm $\mathscr{A}$ is applied at each $x_j \in \mathbb{Z}_N$. Let N_i ($i = 1, \ldots, \ell$) be the number of points $x_j \in \mathbb{Z}_N$ belonging to A_i; that is, N_i is the number of descents to $x_*^{(i)}$ from the points in $\mathbb{Z}_N$. By the definition, $N_i \geqslant 0$ ($i = 1, \ldots, \ell$), $\sum_{i=1}^{\ell} N_i = N$, and the random vector $(N_1, \ldots, N_\ell)$ follow the multinomial distribution

$$\Pr\{N_1 = n_1, \ldots, N_\ell = n_\ell\} = \frac{N!}{n_1! \ldots n_\ell!} \theta_1^{n_1} \ldots \theta_\ell^{n_\ell},$$

where $n_i \geqslant 0$ ($i = 1, \ldots, \ell$) and $\sum_{i=1}^{\ell} n_i = N$.

If ℓ is known, then the problem of drawing statistical inferences concerning the number of local minimizers ℓ, the parameter vector $\theta = (\theta_1, \ldots, \theta_\ell)$ and the number N_* of trials that guarantees with a given probability that all local minimizers are found, is the standard problem of making statistical inferences about parameters of a multinomial distribution. This problem is well documented in literature. The main difficulty is caused by the fact that ℓ is usually unknown. If an upper bound for ℓ is known, then one can apply standard statistical methods; if an upper bound for ℓ is unknown, the Bayesian approach is a natural alternative. Both of these methods are comprehensively studied and explained in [54].

In relation to the algorithm METOD briefly introduced above, we are more interested in whether we have succeeded in finding the global minimizer taking into account that some of our decisions about early termination of local descents may be erroneous. Assume that vol($\mathscr{X}$) $= 1$ and vol(A^*) $= \alpha > 0$, where A^* is the region of attraction of the global minimizer (the value of α does not have to be known, of course). The local descents, including the ones which would lead to the global minimizer, can be stopped early and the corresponding initial points assigned to a wrong region of attraction. The probability of this event is related to the number of checks of the inequality (3.1.4) and the degree of local non-convexity of the objective function. The probability of the fact that an initial point is assigned to a wrong region of attraction is very roughly $1/2^k$, where k is the number of checks of the inequalities (3.1.4) (numerical results show that this probability is much smaller than $1/2^k$, especially for large d).

So we end up with the following rather simple situation. We have a Bernoulli trial with success probability α (when our uniformly distributed starting point x_n belongs to A^*), but on top of this, we have a dropout event (happening with rather small probability κ, which we for simplicity assume the same for all decisions) where we will reassign this initial point to another region of attraction. Therefore, each starting

Table 3.1 Values of $N_{\delta,\gamma}$ for $\gamma = 0.01$ and 0.05, $\delta = 10^{-r}$, $r = 1, 2, \ldots, 8$.

	$r = 1$	$r = 2$	$r = 3$	$r = 4$	$r = 5$	$r = 6$	$r = 7$	$r = 8$
$\gamma = 0.05$	29	299	2995	29956	299572	2995731	29957322	299573226
$\gamma = 0.01$	44	459	4603	46050	460515	4605168	46051700	460517017

point x_n (taken randomly and independently of the other points) will be assigned to A^* with probability at least $\delta = \alpha(1 - \kappa)$. The first starting point assigned to A^* will create a full local descent trajectory converging to the global minimizer. Note that after finding the first point in A^*, the probability of assigning starting points to A^* will increase from $\delta = \alpha(1 - \kappa)$ to $\delta' = \alpha(1 - \kappa) + (1 - \alpha)\kappa/(L - 1)$, where L is the number of local minimizers found so far, as there appears a new possibility of assigning points to A^* when they do not belong there. We can ignore this as we are only interested in the event that at least one initial point will be assigned to A^* and hence that the global minimizer is found.

With N initial i.i.d. uniform starting points, the probability of finding the global minimum is $p_{\delta,N} = 1 - (1 - \delta)^N$. Let $N_{\delta,\gamma}$ be the smallest N such that $p_{\delta,N} \geq 1 - \gamma$; that is, if we choose $N \geq N_{\delta,\gamma}$ then we would guarantee that the probability of finding the global minimizer is at least $1 - \gamma$. Solving the inequality $p_{\delta,N} \geq 1 - \gamma$ with respect to N we find $N_{\delta,\gamma} = \lceil \log \gamma / \log(1 - \delta) \rceil$. Table 3.1 shows some values of $N_{\delta,\gamma}$. From this table we can conclude that there is very little hope of finding the global minimum if the volume of the region attraction of the global minimizer is smaller than 0.00001. On the other hand, if $\delta \leq 0.001$ then METOD of [56] would not require many starting points for guaranteeing high probability of finding the global minimizer.

3.1.3 Convergence and Rate of Convergence

In this section we illustrate the following two features of GRS:

(i) *it is very easy to construct a GRS algorithm which has the theoretical property of convergence;*
(ii) *it is generally impossible to guarantee, even for moderate dimensions, that the global minimum is found in the worst-case scenario.*

Convergence Consider a general GRS algorithm defined by a sequence of probability distributions P_j, $j = 1, 2, \ldots$. We say that this algorithm converges if for any $\delta > 0$, the sequence of points x_j arrives at the set $W(\delta) = \{x \in \mathcal{X} : f(x) - f_* \leq \delta\}$ with probability one. If the objective function is evaluated without error, then this obviously implies convergence (as $n \to \infty$) of record values $y_{0,j} = \min_{i=1\ldots j} f(x_i)$ to f_* with probability 1.

Conditions on the distributions P_j ($j = 1, 2, \ldots$) ensuring convergence of the GRS algorithms are well understood; see, for example, [33, 41]. The results on convergence of GRS algorithms are usually formulated in the form of the 'zero-one law', which is classical in probability theory. The following theorem provides an illustration of such results in a very general setup and is proved in [51, Sect. 3.2] in a more general form.

Theorem 3.2 *Consider a GRS algorithm with $x_j \sim P_j$ for the minimization problem, where $\mathscr{X}$ is a compact set and $f(\cdot)$ is a function on $\mathscr{X}$ satisfying the Lipschitz condition. Let $B(x, \varepsilon) = \{z \in \mathscr{X} : \|z - x\| \leqslant \varepsilon\}$ be a ball centred at x. Define $q_j(\varepsilon) = \inf P_j(B(x, \varepsilon))$, where the infimum is taken over all $x \in \mathscr{X}$, all possible points $x_1, \ldots, x_{j-1}$ and all evaluations of the objective function at these points. Assume that*

$$\sum_{j=1}^{\infty} q_j(\varepsilon) = \infty \tag{3.1.5}$$

for any $\varepsilon > 0$. Then for any $\delta > 0$, the sequence of points x_j falls infinitely often into the set $W(\delta)$, with probability one.

Note that Theorem 3.2 holds in the very general case where evaluations of the objective function $f(\cdot)$ are noisy and the noise is not necessarily random. If the function evaluations are noise-free, then the conditions of Theorem 3.2 ensure that the corresponding algorithm converges; that is, that the sequence of records $y_{0,j}$ converges to f_* with probability 1 and the corresponding subsequence of points $\{x_{i_j}\}$ (where the new records are attained) of the sequence $\{x_j\}$ converges (with probability 1) to the set $\mathscr{X}_* = \{x \in \mathscr{X} : f(x) = f_*\}$ of global minimizers.

If the objective function is evaluated with random error, then the algorithm of generation of points x_j should be accompanied with an algorithm of estimation of the objective function estimation; see [51, Sect. 4.1.3]. Then the rate of convergence of the corresponding algorithm will also depend on the smoothness of the objective function and the chosen approximation routine.

If we use PRS with the uniform distribution $P = P_U$ on $\mathscr{X}$, we obtain $q_j(\varepsilon) = \text{const} > 0$ and therefore the condition (3.1.5) trivially holds. In practice, a usual choice of the distribution P_j is

$$P_j = \alpha_j P_U + (1 - \alpha_j) Q_j, \tag{3.1.6}$$

where $0 \leqslant \alpha_j \leqslant 1$ and Q_j is a specific probability measure on $\mathscr{X}$ which may depend on evaluations of the objective function at the points $x_1, \ldots, x_{j-1}$. Sampling from the distribution (3.1.6) corresponds to taking a uniformly distributed random point in $\mathscr{X}$ with probability α_j and sampling from Q_j with probability $1 - \alpha_j$.

Note that $\sum_{j=1}^{\infty} \alpha_j = \infty$ yields the fulfilment of (3.1.5) for the distributions P_j of the form (3.1.6) and therefore the GRS algorithm with such P_j is theoretically converging. On the other hand, if $\sum_{j=1}^{\infty} \alpha_j < \infty$, then there is a non-zero probability that the neighbourhood of the global minimizer will never be reached.

Unless smoothness conditions about $f(\cdot)$ like the Lipschitz condition are imposed, the statements like Theorem 3.2 are the only tools which are ensuring convergence of the GRS algorithms.

Rate of convergence of PRS Consider a PRS algorithm with $x_j \sim P$. Let $\varepsilon, \delta > 0$ be fixed and B be the target set we want to hit by points x_j. For example, we set $B = W(\delta) = \{x \in \mathscr{X} : f(x) - f_* \leqslant \delta\}$ in the case when the accuracy is expressed in terms of closeness with respect to the function value and $B = B(x_*, \varepsilon)$ if we are studying convergence towards the global minimizer x_*. Assume that P is such that $P(B) > 0$; for example, $P = P_U$ is the uniform probability measure on $\mathscr{X}$.

Define the Bernoulli trials where the success in the trial j mean that $x_j \in B$. PRS generates a sequence of independent Bernoulli trials with the same success probability $\Pr\{x_j \in B\} = P(B)$. In view of the independence of x_j, we have $\Pr\{x_1 \notin B, \ldots, x_n \notin B\} = (1 - P(B))^n$, and therefore the probability

$$\Pr\{x_j \in B \text{ for at least one } j, \ 1 \leqslant j \leqslant n\} = 1 - (1 - P(B))^n$$

tends to one as $n \to \infty$. We also assume that $P(B)$ is small.

Let n_γ be the number of points which are required for PRS to reach the set B with probability at least $1 - \gamma$, where $\gamma \in (0, 1)$; that is, $n_\gamma = \min\{n : 1 - (1 - P(B))^n \geqslant 1 - \gamma\}$. Solving it we obtain

$$n_\gamma = \lceil \ln \gamma / \ln (1 - P(B)) \rceil \cong (-\ln \gamma)/P(B)$$

as $P(B)$ is small and $\ln (1 - P(B)) \cong -P(B)$ for small $P(B)$.

The numerator in the expression for n_γ depends on γ, but it is not large; for example, $-\ln \gamma \simeq 2.996$ for $\gamma = 0.05$. But the denominator, which is approximately $P(B)$, can be extremely small and hence n_γ could be astronomically large.

Consider an example where $\mathscr{X} = [0, 1]^d$, $P = P_U$ on $\mathscr{X}$ and $B = B(x_*, \varepsilon)$. Then $P(B) = \text{vol}(B) \leqslant \varepsilon^d V_d$, where $V_d = \pi^{\frac{d}{2}}/\Gamma\left(\frac{d}{2} + 1\right)$ is the volume of the unit ball in $\mathbb{R}^d$, where $\Gamma(\cdot)$ is the gamma function. In view of the upper bound of the form 'const $\cdot \varepsilon^d$', the probability $P(B)$ can be extremely small even when ε is not very small (say, $\varepsilon = 0.1$) and $d \geqslant 10$. The number n_γ in this case is illustrated in Fig. 3.2.

Rate of convergence of a general GRS method The easiest way to ensure convergence of a general GRS algorithm is to choose the probabilities P_j in the form (3.1.6), where the coefficients α_j satisfy the condition (3.1.5); see Sect. 3.1.3.

Let us generalize the arguments given in Sect. 3.1.3 for the case of PRS, to the case of a general algorithm of GRS. Instead of the equality $\Pr\{x_j \in B\} = P(B)$ for all $j \geqslant 1$, we now have the inequality $\Pr\{x_j \in B\} \geqslant \alpha_j P_U(B)$, where the equality holds in the worst-case scenario. We define $n(\gamma)$ as the smallest integer such that the inequality $\sum_{j=1}^{n(\gamma)} \alpha_j \geqslant -\ln \gamma / P_U(B)$ is satisfied. For the choice $\alpha_j = 1/j$, which is a common recommendation, we can use the approximation $\sum_{j=1}^{n} \alpha_j \simeq \ln n$.

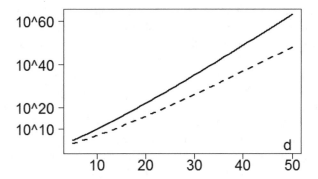

Fig. 3.2 The number n_γ of points which are required for PRS to reach the set $B = B(x_*, \varepsilon)$ with probability at least $1 - \gamma = 0.95$ for $\varepsilon = 0.1$ (solid) and $\varepsilon = 0.2$ (dashed) as the dimension d varies in $[5, 50]$

Fig. 3.3 The number $n(\gamma)$ of points which are required for GRS to reach the set $B = B(x_*, \varepsilon)$ with probability at least $1 - \gamma = 0.95$ for $\varepsilon = 0.1$ (solid) and $\varepsilon = 0.2$ (dashed) as the dimension d varies in $[5, 50]$

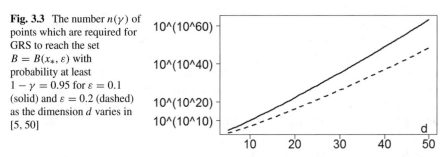

Therefore we obtain $n(\gamma) \simeq \exp\{-\ln \gamma / P_U(B)\}$. For the case of $\mathscr{X} = [0, 1]^d$ and $B = B(x_*, \varepsilon)$, we obtain $n(\gamma) \simeq \exp\{c \cdot \varepsilon^{-d}\}$, where $c = (-\ln \gamma)/V_d$. Note also that if the distance between x_* and the boundary of $\mathscr{X}$ is smaller than ε, then the constant c and hence $n(\gamma)$ are even larger. For example, for $\gamma = 0.1$, $d = 10$ and $\varepsilon = 0.1$, $n(\gamma)$ is larger than $10^{1000000000}$. Even for optimization problems in a small dimension $d = 3$, and for $\gamma = 0.1$ and $\varepsilon = 0.1$, the number $n(\gamma)$ of points required for the GRS algorithm to hit the set B in the worst-case scenario is huge, namely, $n(\gamma) \simeq 10^{238}$. Figure 3.3 shows the behaviour of $n(\gamma)$ as the dimension grows.

The main conclusion of this discussion is as follows: *even for moderate dimensions, general GRS algorithms do not guarantee convergence in practical computations.* Convergence could only be seriously discussed if the Lipschitz-type conditions are assumed and used in the process of search.

3.1.4   Meta-heuristics in GRS

Heuristics are problem-dependent techniques and are usually adapted to the problem at hand by trying to take full advantage of the specificity of the problem. Meta-heuristics are techniques that can be used for a large class of objective functions which can be treated as black boxes. The meta-heuristics involving randomized decisions (which most of them do) and designed to search for the global optimum (rather than just local one) should therefore be considered as components of the subject which we call *Methodology of GRS*.

There exists a large number of different meta-heuristics; see, e.g. [1, 3, 10, 12, 19, 37]. There are hundreds of names for the corresponding algorithms including 'imperialist competitive', 'shuffled frog leaping', 'intelligent water drops', 'flower pollination' as well as many other extravagant names. There are many specialized journals on meta-heuristics and evolutionary algorithms with large impact factors. The number of citations to many papers on meta-heuristics are in the hundreds and even thousands. Of course, the main reason for such popularity of meta-heuristics is usefulness of such techniques for solving practical problems, especially in discrete optimization (which we are not discussing in this chapter). One cannot solve a complicated practical problem by constructing GRS algorithms based solely on theoretical grounds and basic methodological principles (the approach chosen in [51, 53] and in this chapter), and in this sense, the approach I have followed all my scientific life is rather restrictive. Also, theoretical considerations alone cannot inspire fertile ideas such as [7]. On the other hand, development of meta-heuristics without understanding basic methodological principles of GRS may be counter-productive. Examples of two serious misunderstandings by the meta-heuristic community are the huge popularity of (a) basic simulated annealing [47] and (b) basic genetic algorithms [20, 36] in continuous optimization problems. I have been emphasizing the misunderstandings and theoretical flaws in these methods for more than 30 years [49, 51] but did not obtain a lot of audience. I will summarize the arguments against basic simulated annealing and basic genetic algorithms; see Sect. 3.1.1 and the end of Sect. 3.3.2, respectively.

My view on the field of meta-heuristics in GRS is as follows. First, there are many positive features such as (a) many very intelligent people are working in the field; (b) there are many examples of very successful applications of meta-heuristics in practice; and (c) there are many user-friendly packages available for common use. However, it is often difficult to immediately see the difference between good and bad meta-heuristics as all algorithms have many parameters to be tuned and comparative testing of algorithms is notoriously hard; see e.g. [23] and [12, Ch.21]. In view of the lack of theory (in most cases), this yields that it is relatively easy to publish a paper with meta-heuristical arguments in a journal with high impact factor (unfortunately, this is one of the major formal objectives in modern organization of scientific community). As writing such a paper does not necessarily imply that the author of the paper has sufficient educational level and is even aware of existence of scientific literature on the subject of stochastic global optimization, the meta-heuristics community arrived at a sad situation where good publications are often lost in-between very large number of very poor, conceptually wrong, misleading and confusing publications.

3.2 Statistical Inference About f_*

Let $\mathscr{X}$ be a compact in $\mathbb{R}^d$ and $x_1, \ldots, x_n$ be points constructed by PRS with $x_j \sim P$, where n is a large number and P is a probability measure on $\mathscr{X}$ with some density $p(x)$, which is a piece-wise continuous function on $\mathscr{X}$ and $p(x) > 0$ for all $x \in \mathscr{X}$. Using prior information about $f(\cdot)$ and considering the values $\{f(x_j)\}_{j=1,\ldots,n}$ as a sample, we can make statistical inference of the following types: (1) building either a parametric or non-parametric estimator of $f(\cdot)$, e.g. a Bayesian estimator or an estimator based on the Lipschitz condition, and (2) construction of an estimator and a confidence interval for f_*. Below we only consider inferences of type (2) following an approach comprehensively discussed in [51, Ch. 7]. Statistical inference about f_*, the minimal value of the objective function $f(\cdot)$, can serve, for example, for the following purposes: (i) devising specific GRS algorithms like the branch and probability bounds methods (see [50, 55] and [51, Sect. 4.3]), (ii) constructing stopping rules (see [52]) and (iii) increasing efficiency of the population-based GRS methods.

Any population-based GRS method (see Sect. 3.3) can be enhanced with such statistical procedures, which could be useful for (a) helping to make decisions when to stop creating the current population and start making a new one, (b) constructing the overall stopping rule and (c) improving decision rules for deciding on prospectiveness of different subsets of $\mathscr{X}$. In this respect, we believe that the technique developed in [52] has a serious potential. Another example is the use of statistical inferences in GRS algorithms for solving multiobjective optimization problems with non-convex objectives as proposed in [55]. Possibilities are enormous; in particular, for those who like meta-heuristical GRS algorithms, see a recent paper [26] for an example of such algorithm.

3.2.1 Statistical Inference in PRS: The Main Assumption

Since the points x_j in PRS are i.i.d. with distribution P, the elements of the sample $Y = \{y_1, \ldots, y_n\}$ with $y_j = f(x_j)$ are i.i.d. with c.d.f.

$$F(t) = \Pr\{x \in \mathscr{X} : f(x) \leqslant t\} = \int_{f(x) \leqslant t} P(dx) = P(W(t - f_*)), \qquad (3.2.1)$$

where $t \geqslant f_*$ and $W(\delta) = \{x \in \mathscr{X} : f(x) \leqslant f_* + \delta\}$, $\delta \geqslant 0$. Note that the c.d.f. $F(t)$ is concentrated on the interval $[f_*, f^*]$, where $f^* = \max_{x \in \mathscr{X}} f(x)$, and our main interest is the unknown value f_*, which is the lower bound of this interval. Since the analytic form of $F(t)$ is either unknown or incomprehensible (unless f is very simple), for making statistical inferences about f_*, we need to use asymptotic considerations based the record values of the sample Y. It is well-known (see, e.g. [27]) that the asymptotic distribution of the order statistics is unambiguous and the conditions on $F(t)$ and $f(\cdot)$, when this asymptotic law works, are very mild and typically hold in real-life problems. Specifically, for a very wide class of functions $f(\cdot)$ and distributions P, the c.d.f. $F(t)$ can be represented as

$$F(t) = c_0(t - f_*)^\alpha + o((t - f_*)^\alpha), \ t \downarrow f_*, \qquad (3.2.2)$$

where c_0 and α are some positive constants. Moreover, the coefficient $c_0 = c_0(t)$ can be a slowly varying function for $t \simeq f_*$, and the results given below are also valid for this more general case. In our constructions the value of c_0 is not important, but the value of α is critical. The coefficient α is called 'tail index', and its value is often known, as discussed in Sect. 3.2.2.

Denote by η a random variable which has c.d.f. (3.2.1) and by $y_{1,n} \leqslant \ldots \leqslant y_{n,n}$ the order statistics corresponding to the sample Y. Note that f_* is the lower endpoint of the random variable η. One of the fundamental results in the theory of extreme order statistics states (see, e.g. [27]) that if (3.2.2) holds, then $F(t)$ belongs to the domain of attraction of the Weibull distribution with density $\psi_\alpha(t) = \alpha t^{\alpha-1} \exp\{-t^\alpha\}$, $t > 0$. This distribution has only one parameter, the tail index α.

3.2.2 Tail Index

As stated in Sect. 3.2.1, the representation (3.2.2) holds in most real-life problems. The main issue is whether the value of the tail index α can be specified or has to be estimated. The second option, that is the estimation of α, is very difficult; see, e.g. [6] for a survey on comparison of different estimators of α. The number n of points must be astronomically large (even for small dimension d) if we want to accurately estimate f_* after replacing α with any estimator. Practically, n should be extremely large to see any convergence of estimators; see discussion in [53, Sect. 2.5.1]. Asymptotically, as $n \to \infty$, if α is estimated, then the asymptotic mean squared error (MSE) of the maximum likelihood estimator (MLE) of f_* is at least $(\alpha - 1)^2$ times larger than the MSE of the MLE of f_* in the case when α is known.

In PRS we can usually have enough knowledge about $f(\cdot)$ to get the exact value of the tail index α. In particular, the following results hold: if the global minimizer x_* of $f(\cdot)$ is unique and $f(\cdot)$ is locally quadratic around x_*, then the representation (3.2.2) holds with $\alpha = d/2$. Moreover, if the global minimizer x_* of $f(\cdot)$ is attained at the boundary of $\mathscr{X}$ and the gradient of $f(\cdot)$ has all non-zero components at x_*, then the representation (3.2.2) holds with $\alpha = d$. For a detailed exposition of the related theory, generalizations and history of this result, see [51].

The fact that α has the same order as d, when d is large, implies 'the curse of dimensionality'. We theoretically study this in the following sections, but in this section we illustrate this curse of dimensionality on a simple numerical example.

Consider the minimization problem with the objective function $f(x) = e_1^T x$, where $e_1 = (1, 0, \ldots, 0)^T$ and the set $\mathscr{X}$ is the unit ball: $\mathscr{X} = \{x \in \mathbb{R}^d : ||x|| \leqslant$

1}. It is easy to see that the minimal value is $f_* = -1$ and the global minimizer $z_* = (-1, 0, \ldots, 0)^T$ is located at the boundary of $\mathscr{X}$. Consider PRS with points $x_j \sim P_U$. In Fig. 1.5 we have depicted projections of points $x_1, \ldots, x_n$ on a fixed two-dimensional plane for $n = 10^3$ and $n = 10^5$ and the dimension $d = 20$. Even if the number of simulated points is large, we can observe a thick ring inside the unit circle with no projections of points although the points are uniformly distributed in the unit hyperball.

Define $r_j = ||x_j||$. It is well-known that $\Pr(r_j < t) = t^d$. Thus, the distribution of r_j satisfies the representation (3.2.2) with $\alpha = d$. We are interested in the record values for the sample with $y_j = e_1^T x_j$, $j = 1, \ldots, n$.

Let us give some numerical values. In a simulation with $n = 10^3$ and $d = 20$, we have received $y_{1,n} = -0.64352$, $y_{2,n} = -0.61074$, $y_{3,n} = -0.60479$ and $y_{4,n} = -0.60208$. In a simulation with $n = 10^5$ and $d = 20$, we have obtained $y_{1,n} = -0.74366$, $y_{2,n} = -0.73894$, $y_{3,n} = -0.73228$ and $y_{4,n} = -0.72603$. In Fig. 3.4 we depict the differences $y_{k,n} - f_*$ for $k = 1, 4, 10$ and $n = 10^3, \ldots, 10^{13}$, where the horizontal axis has logarithmic scale. We can clearly see that the difference $y_{k,n} - y_{1,n}$ is much smaller than the difference $y_{1,n} - f_*$; that shows that the problem of estimating the minimal value of f_* is very difficult.

In Fig. 3.5 we show that the difference $y_{1,n} - f_*$ increases as the dimension d grows, for fixed n. Thus, the minimization problem becomes harder in larger dimensions. Also, Fig. 3.5 shows that difference $y_{10,n} - y_{1,n}$ is much smaller than the difference $y_{1,n} - f_*$.

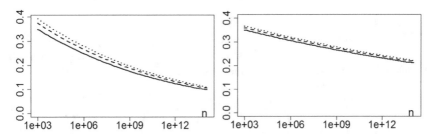

Fig. 3.4 Differences $y_{1,n} - f_*$ (solid), $y_{4,n} - f_*$ (dashed) and $y_{10,n} - f_*$ (dotted), where $y_{k,n}$, $k = 1, 4, 10$, are records of evaluations of the function $f(x) = e_1^T x$ at points $x_1, \ldots, x_n$ with uniform distribution in the unit hyperball in the dimension $d = 20$ (left) and $d = 50$ (right)

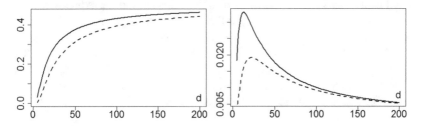

Fig. 3.5 The difference $y_{1,n} - f_*$ (left) and $y_{10,n} - y_{1,n}$ (right) for $n = 10^6$ (solid) and $n = 10^{10}$ (dashed), where $y_{j,n}$ is the j-th record of evaluations of the function $f(x) = e_1^T x$ at points $x_1, \ldots, x_n$ with uniform distribution in the unit hyperball in the dimension d; d varies in [5, 200]

3.2.3   Estimation of the Minimal Value of f

In this section, we review the asymptotic properties of two estimators of the minimal value f_*, the MLE and the best linear estimator. We also discuss properties of these estimators in the case of large dimension d (and hence large α).

If the representation (3.2.2) holds, $\alpha \geq 2$ is fixed, $k \to \infty$, $k/n \to 0$ as $n \to \infty$, then the MLE $\hat{f}_{mle}$ of f_* is asymptotically normal and asymptotically efficient in the class of asymptotically normal estimators, and the MSE has the asymptotic form

$$E(\hat{f}_{mle} - f_*)^2 \approx \begin{cases} (1 - \frac{2}{\alpha})(\kappa_n - f_*)^2 k^{-1+2/\alpha}, & \alpha > 2, \\ (\kappa_n - f_*)^2 / \ln k, & \alpha = 2, \end{cases} \qquad (3.2.3)$$

where κ_n is the $(1/n)$-quantile of the c.d.f. $F(\cdot)$. In view of (3.2.2) we have

$$\kappa_n - f_* = (c_0 n)^{-1/\alpha} (1 + o(1)) \quad \text{as } n \to \infty. \qquad (3.2.4)$$

Linear estimators of f_* are simpler than the MLE. Nevertheless, the best linear estimators have the same asymptotic properties. To define a linear estimator, we introduce the vectors $a = (a_1, \ldots, a_k)^T \in \mathbb{R}^k$, $\mathbf{1} = (1, 1, \ldots, 1)^T \in \mathbb{R}^k$, $b = (b_1, \ldots, b_k)^T \in \mathbb{R}^k$, where $b_i = \Gamma(i+1/\alpha) / \Gamma(i)$, and the matrix $\Lambda = \|\lambda_{ij}\|_{i,j=1}^k$, where $\lambda_{ji} = \lambda_{ij} = u_i v_j$, $i \geq j$, and

$$u_i = \Gamma(i+2/\alpha)/\Gamma(i+1/\alpha), \quad v_j = \Gamma(j+1/\alpha)/\Gamma(j).$$

The matrix Λ in such form can be inverted analytically.

A general linear estimator of f_* can be written as $\hat{f}_{n,k}(a) = \sum_{i=1}^k a_i y_{i,n}$, where $a = (a_1, \ldots, a_k)^T$ is the vector of coefficients. Then using explicit expressions for moments of order statistics, for any linear estimator $\hat{f}_{n,k}(a)$, we obtain

$$E \hat{f}_{n,k}(a) = \sum_{i=1}^k a_i E y_{i,n} = f_* \sum_{i=1}^k a_i + (\kappa_n - f_*) a^T b + o(\kappa_n - f_*), \quad n \to \infty.$$

Since $\kappa_n - f_* \to 0$ as $n \to \infty$ and the variances of all $y_{i,n}$ are finite, the estimator $\hat{f}_{n,k}(a)$ is a consistent estimator of f_* iff $a^T \mathbf{1} = \sum_{i=1}^k a_i = 1$. Using explicit expressions for the moments of order statistics and the expression (3.2.4), we obtain the following expression for the MSE of the estimator $\hat{f}_{n,k}(a)$:

$$E(\hat{f}_{n,k}(a) - f_*)^2 = (c_0 n)^{-2/\alpha} a^T \Lambda a (1 + o(1)), \quad n \to \infty. \qquad (3.2.5)$$

The asymptotic MSE (3.2.5) is a natural optimality criterion for choosing the vector of coefficients a, whose minimum is attained at

$$a^* = \arg \min_{a: a^T \mathbf{1}=1} a^T \Lambda a = \Lambda^{-1} \mathbf{1} / \mathbf{1}^T \Lambda^{-1} \mathbf{1}, \qquad (3.2.6)$$

with $\min_{a:a^T\mathbf{1}=1} a^T \Lambda a = (a^*)^T \Lambda a^* = 1/\mathbf{1}^T \Lambda^{-1}\mathbf{1}$. The estimator $\hat{f}_{n,k}(a^*)$ is called the optimal linear estimator; it has been proposed in [5], where the form (3.2.6) was obtained.

As shown in [51, Th. 7.3.2], the components of the vector $a^* = (a_1^*, \ldots, a_k^*)^T$ can be expressed explicitly as follows, $a_i^* = v_i/\mathbf{1}^T \Lambda^{-1}\mathbf{1}$ for $i = 1, \ldots, k$, where

$$
\begin{aligned}
v_1 &= (\alpha + 1)/\Gamma(1 + 2/\alpha), \\
v_i &= (\alpha - 1)\Gamma(i)/\Gamma(i + 2/\alpha), \qquad i = 2, \ldots, k - 1, \\
v_k &= -(\alpha k - \alpha + 1)\Gamma(k)/\Gamma(k + 2/\alpha).
\end{aligned}
$$

and

$$
\mathbf{1}^T \Lambda^{-1}\mathbf{1} = \begin{cases} \frac{1}{\alpha-2}\left(\frac{\alpha\Gamma(k+1)}{\Gamma(k+2/\alpha)} - \frac{2}{\Gamma(1+2/\alpha)}\right), & \alpha \neq 2, \\ \sum_{i=1}^k 1/i, & \alpha = 2. \end{cases} \tag{3.2.7}
$$

Note that the expression (3.2.7) is valid for all $\alpha > 0$ and $k = 1, 2, \ldots$. Using the Taylor series $\Gamma(k + 2/\alpha) = \Gamma(k) + \frac{2}{\alpha}\Gamma'(k) + O(1/\alpha^2)$ for large values of α, we obtain

$$
\min_{a:a^T\mathbf{1}=1} a^T \Lambda a = \frac{1}{\mathbf{1}^T \Lambda^{-1}\mathbf{1}} \simeq \frac{1}{k} + \frac{2(\psi(k) - 1 + 1/k)}{\alpha k}, \tag{3.2.8}
$$

for large α, where $\psi(\cdot) = \Gamma'(\cdot)/\Gamma(\cdot)$ is the psi-function. In view of (3.2.5), the coefficient $(a^*)^T \Lambda a^*$ in the expression for the MSE of the optimal linear estimator $\hat{f}_{n,k}(a^*)$ is nearly constant for large dimensions and has little effect on the rate of convergence of the MSE of $\hat{f}_{n,k}(a^*)$. The quality of approximation (3.2.8) is illustrated in Figs. 3.6 and 3.7.

The asymptotic properties (when both n and k are large) of the optimal linear estimators coincide with the properties of the MLE and hold under the same regularity conditions, as proved in [51, Sect. 7.3.3]. In particular, the optimal linear estimator $\hat{f}_{n,k}(a^*)$ is asymptotically normal (as $n \to \infty$, $k \to \infty$, $k/n \to 0$) and the mean square error $E(\hat{f}_{n,k}(a^*) - f_*)^2$ asymptotically behaves like (3.2.3).

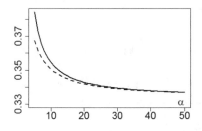 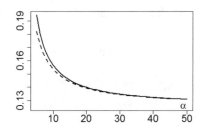

Fig. 3.6 The exact expression of $1/\mathbf{1}^T \Lambda^{-1}\mathbf{1}$ (solid) and the approximation (3.2.8) (dashed) for $k = 3$ (left) and $k = 8$ (right); α varies in [5, 50]

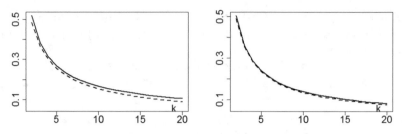

Fig. 3.7 The exact expression of $1/\mathbf{1}^T \Lambda^{-1} \mathbf{1}$ (solid) and the approximation (3.2.8) (dashed) for $\alpha = 5$ (left) and $\alpha = 8$ (right); as k varies in [2, 20]

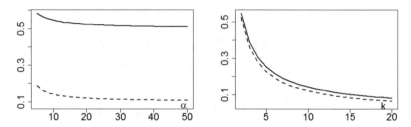

Fig. 3.8 Asymptotic efficiency (3.2.9) of $y_{1,n}$. Left: $k = 2$ (solid) and $k = 10$ (dashed); as α varies in [5, 50]. Right: $\alpha = 10$ (solid) and $\alpha = 20$ (dashed); as k varies in [2, 20]

Note that the efficiency of the estimators $\hat{f}_{mle}$ and $\hat{f}_{n,k}(a^*)$ can be very low if an incorrect value of α is used for computing this estimator; see [53, Sect. 2.5.2]. Even if the correct value of α is known, all estimators including $\hat{f}_{mle}$ and $\hat{f}_{n,k}(a^*)$ may be severely biased. It is caused by the difficulty of reaching the neighbourhood where the asymptotic equality (3.2.2) holds. This is the case, in particular, for the problem of estimating the covering radius studied in Sect. 1.3; see [28].

In practice of global optimization, the standard estimator of f_* is the current record $y_{1,n}$, which can be written as $y_{1,n} = \hat{f}_{n,k}(e_1)$ where $e_1 = (1, 0, 0, \ldots, 0)^T$. By (3.2.5), the MSE of $y_{1,n}$ is

$$E(\hat{f}_{n,k}(e_1) - f_*)^2 = \Gamma(1 + 2/\alpha)(c_0 n)^{-2/\alpha} \left(1 + o(1)\right), \quad n \to \infty.$$

Asymptotic efficiency of $y_{1,n}$ is therefore

$$\text{eff}(y_{1,n}) = \left[\Gamma(1 + 2/\alpha) \cdot \mathbf{1}^T \Lambda^{-1} \mathbf{1} \right]^{-1}. \tag{3.2.9}$$

In view of (3.2.8), this efficiency tends to $1/k$ if k is fixed and $\alpha \to \infty$. The asymptotic behaviour of the efficiency (3.2.9) is illustrated in Fig. 3.8.

3.2.4 Comparison of Random and Quasi-random Sequences

There are many attractive features of good GRS methods, but is it so important to choose points x_j at random? Rephrasing this, can we improve efficiency of GRS algorithms if we sacrifice some randomness? The answer to this question is similar to what you may find in other areas of applied mathematics like Monte Carlo methods for estimation of integrals: with randomness, we gain simplicity of the methods and a possibility to make statistical inferences but usually lose some efficiency.

First, we need to perform local descents (which are parts of many GRS algorithms) using standard deterministic routines like conjugate gradient method; indeed, local random search algorithms cannot compete with such methods. Second, during the exploration stage of GRS methods, purely deterministic (e.g. quasi-random) sequences of points would do this exploration much more efficiently than random sequences. If in place of random points in PRS we use any of the quasi-random sequences (say, a low-dispersion sequence), then we will (i) improve the rate of convergence of PRS investigated in Sect. 3.1.3, (ii) avoid very long waiting times with infinite expectation for getting new records and (iii) gain reproducibility of results. If we use some semi-random sequences like the stratified sample in place of i.i.d. sample in the PRS, then we still will be able to use some of the statistical procedures outlined above. More precisely, consider a version of PRS where the sample $\{x_1, \ldots, x_n\}$ is stratified rather than independent. Assume that the distribution $P = P_U$ is uniform on $\mathscr{X}$ and the set $\mathscr{X}$ is split into m subsets of equal volume. Assume also that in each subset we generate l independent uniformly distributed points. The sample size is then $n = ml$. In particular, under the assumption $l > k$ and similar assumptions about $f(\cdot)$, the same asymptotically optimal estimator with coefficients (3.2.6) can be used. Its accuracy is better than the accuracy of the same estimator computed for independent sample; see [51] Sect. 3.2.

GRS algorithms compared with deterministic optimization procedures have a very attractive feature: in GRS algorithms we can use statistical procedures for increasing efficiency of the algorithms and devising stopping rules. But do we gain much by choosing the points at random and can we improve the efficiency of GRS algorithms if we sacrifice some randomness? The answer to this question is similar to what we know from other areas of applied mathematics like estimation of integrals using Monte Carlo methods and cubature formulas: randomness provides simplicity of methods and possibility of making statistical inferences; however, for small dimensions we can significantly improve efficiency by reducing randomness and making the best possible deterministic decisions.

One possible recommendation for combining random search and deterministic procedures can be formulated as follows. First of all, if a global optimization method requires local descents, then for doing, it is advisable to use standard deterministic routines like the conjugate gradient method (since local random search algorithms would never be able to compete with such methods). In the stage, where a GRS algorithm explores the whole $\mathscr{X}$ or some prospective subsets of $\mathscr{X}$, purely deterministic or quasi-random sequences of points may do this exploration more efficiently than random sequences, especially in small dimensions. If PRS will use any of the quasi-random sequence instead of random points, then this will improve the rate of convergence of PRS in low dimensions only, avoid very long waiting times with infinite expectation for getting new records (in the purely random version of PRS) and gain the reproducibility of results. However, if the dimension d is large, then the use of quasi-random points instead of purely random may no longer bring visible advantage. Consider an algorithm of global search in $\mathscr{X} = [0, 1]^d$, where points create a low-dispersion sequence relative to the L_∞-metric in the multidimensional case. For every dimension d and any n-point sequence $\mathbb{Z}_n = \{x_1, \ldots, x_n\}$, the dispersion (with respect to L_∞-metric ρ_∞) $d_\infty(\mathbb{Z}_n) = \max_{x \in \mathscr{X}} \min_{i=1,\ldots,n} \rho_\infty(x, x_i)$ satisfies the inequality $d_\infty(\mathbb{Z}_n) \geq 0.5 n^{-1/d}$, and there exists a sequence $\mathbb{Z}_n^*$ such that $\lim_{n \to \infty} n^{1/d} d_\infty(\mathbb{Z}_n^*) = 1/(2 \ln 2)$. This means that the rate of covering of the set $\mathscr{X}$ by points from the best low-dispersion sequence has the order $O(n^{-1/d})$, which coincides with the rate achieved by PRS with uniform distribution P_U. Numerical results presented in [32] confirm this.

3.3 Population-Based Algorithms

Simulated annealing and other Markovian global random search algorithms make use of some information about the objective function gained during the process of search. Although this information is limited to the last observation only, many Markovian algorithms have proven to be more practically efficient than the non-adaptive search algorithms. Thus, the possibility to use information even at one point leads to a visible improvement in efficiency.

A natural next step in improving the efficiency of global random search algorithms is to allow the possibility of using more information about the objective function keeping the structure of the algorithms relatively simple. The algorithms discussed in this section do exactly this: these are the algorithms transforming one group of points (current generation) to another group of points (next generation) by certain probabilistic rules. We shall call these algorithms 'population-based algorithms'. Note that the family of population-based algorithms includes many evolutionary algorithms including the very popular 'genetic algorithms'.

3.3.1 Construction of Algorithms and Their Convergence

We shall use the general scheme of global random search algorithms presented in the beginning of Sect. 3.1. We assume that the rule for constructing the probability distributions $P_{j+1}(\cdot)$ does not make use of the points $x_{l_i}^{(i)}$ ($l_i = 1, \ldots, n_i$; $i = 1, \ldots, j-1$) and the results of the objective function evaluation at these points.

That is, the probability distributions $P_{j+1}(\cdot)$ are constructed using only the points of the j-th iteration $x_l^{(j)}$ ($l = 1, \ldots, n_j$) and the results of the objective function evaluation at these points. The resulting algorithms will be called 'population-based algorithms'. For convenience of references, let us formulate the general form of these algorithms.

Algorithm 1 (Population-based algorithms: general form)

1. *Choose a probability distribution $P_1(\cdot)$ on the n_1–fold product set $\mathscr{X} \times \ldots \times \mathscr{X}$, where $n_1 \geqslant 1$ is a given integer. Set iteration number $j = 1$.*
2. *Obtain n_j points $x_1^{(j)}, \ldots, x_{n_j}^{(j)}$ in A by sampling from the distribution $P_j(\cdot)$. Evaluate the objective function $f(\cdot)$ at these points.*
3. *Check a stopping criterion.*
4. *Using the points $x_l^{(j)}$ ($l = 1, \ldots, n_j$) and the objective function values at these points, construct a probability distribution $P_{j+1}(\cdot)$ on the n_{j+1}–fold product set $\mathscr{X} \times \ldots \times \mathscr{X}$, where n_{j+1} is some integer that may depend on the search information.*
5. *Substitute $j+1$ for j and return to Step 2.*

We shall call the set of points of the j-th iteration

$$x_1^{(j)}, \ldots, x_{n_j}^{(j)} \tag{3.3.1}$$

the 'parent generation' and the related set of points of the $(j+1)$-th iteration

$$x_1^{(j+1)}, \ldots, x_{n_{j+1}}^{(j+1)} \tag{3.3.2}$$

the 'generation of descendants' or 'children'.

To define a specific population-based algorithm, one has to define
(a) the stopping rule,
(b) the rules for computing the numbers n_j (population sizes), and
(c) the rules for obtaining the population of descendants (3.3.2) from the population of parents (3.3.1).

Stopping rules

– The simplest stopping rule is based on counting the total number of points generated.
– If the objective function satisfies the Lipschitz-type condition, then for defining the stopping rule, one can use the recommendations of [51, Sect.3.1].
– If a local descent is routinely used in obtaining the descendants, then the statistical techniques discussed in Sect. 3.1.2 may be employed.
– If the population sizes n_j are large enough and all the descendants (3.3.2) are generated using the same probabilistic rule, then to devise the stopping rule, one can use the statistical procedures to infer about the minimal value of $f(\cdot)$ in A and in subsets of A.

Choice of the population sizes n_j The population-based algorithms in the form of Algorithm 1 have much more flexibility than the Markovian algorithms in adapting the search for increasing its efficiency.

The choice of population sizes n_j is an important tool in creating efficient population-based algorithms for various classes of optimization problems. Large values of n_j are used for increasing the globality of the search. Small n_j make the search more local. In the limiting case, when for some j_0 we have $n_j = 1$ ($j \geqslant j_0$), the search (after iteration j_0) becomes Markovian, which is almost local.

There are no general restrictions on the ways of choosing the population sizes n_j. We distinguish the following four ways of selecting the values n_j:

(i) n_j are random and depend on the statistical information gathered during the search;
(ii) the sequence of n_j is non-increasing: $n_1 \geqslant n_2 \geqslant \ldots \geqslant n_j \geqslant \ldots$;
(iii) $n_j = n$ for all j;
(iv) the sequence of n_j is non-decreasing: $n_1 \leqslant n_2 \leqslant \ldots \leqslant n_j \leqslant \ldots$

The rule (i) may lead to very efficient algorithms. If one wants to use rule (i), we advice using statistical procedures and appropriate recommendations of Sect. 3.2.

From the practical viewpoint, the choice (ii) seems very natural. Indeed, in the first few iterations of Algorithm 1, we need to make the search more global, and therefore it is normal to choose the first few n_j large. As the search progresses, it is natural to assume that this search reduces the uncertainty about the minimizer(s), and it is therefore natural to narrow the search area. In this way, it may be a good idea to keep reducing the population sizes and pay more and more attention to the local search (to accelerate local convergence to the minimizers).

The choice (iii) is the most convenient from the theoretical point of view. This case is studied in Sect. 3.3.2. Note that the value of n does not have to be large. Even in the case $n = 2$, Algorithm 1 already significantly generalizes the Markovian algorithms; at the same time, its structure is still relatively simple.

Using the choice (iv) does not seem natural. However, it looks like that there is no other way of achieving the convergence of all the points (3.3.1) (as $j \to \infty$) to the neighbourhood of the global minimizer(s) in the case when the objective function is evaluated with random error (this case is thoroughly investigated in [51], Sect. 5.2). Indeed, as long as we approach the global minimizer, we need to diminish the effect of random errors; this can only be done if $n_j \to \infty$ as j increases.

Obtaining the population of descendants from the population of parents   The major distinction between different versions of Algorithm 1 is related to the rules which define the way of obtaining the population of descendants (3.3.2) from the population of parents (3.3.1).

A very convenient way to obtain the population of descendants (3.3.2) is to obtain each descendant separately using the same probabilistic rule (this would imply that the descendants are independent random points conditionally the parent population is given). In many algorithms, each descendant $x_l^{(j+1)}$ has only one parent $x_i^{(j)}$, where $i \in \{1, \ldots, n_j\}$ is typically a number computed with a help of some probabilistic rule (see below for related heuristics). Then to move from the parent $x_i^{(j)}$ to the current descendant $x_l^{(j+1)}$, one has to perform an operation which is called 'mutation' in genetic algorithms. In probabilistic language, it corresponds to sampling some probability distribution $Q_j(x_i^{(j)}, \cdot)$ which is called the transition probability. The transition probability defines the way of choosing the point of the next generation in the neighbourhood of a chosen point from the parent generation.

For the sake of simplicity, the transition probabilities $Q_j(x, \cdot)$ are often chosen so that for sampling $Q_j(x, \cdot)$ one samples uniform distribution on $\mathscr{X}$ with small probability $p_j \geqslant 0$ and some other distribution (which depends on x) with probability $1 - p_j$. For example, this distribution can be the uniform distribution on either a ball or a cube with centre at x and volume depending on j; it can even correspond to performing several iterations of a local descent algorithm starting at x. In either case, the condition $\sum_{j=1}^{\infty} p_j = \infty$ guarantees the convergence of the algorithm; see Sect. 3.1.3.

3.3.2 Homogeneous Transition Probabilities

In the present section, we consider the case where the population sample size is constant (i.e. $n_j = n$ for all j) and the transition probabilities $Q_j(z, dx)$ are time-homogeneous (i.e., $Q_j(z, dx) = Q(z, dx)$ for all j). In this case, the population-based algorithms behave like ordinary Markov chains, and the study of their asymptotic properties is reduced to (i) the study of convergence to the stationary distributions and (ii) the study of properties of these stationary distributions.

Below in this section, we establish (under rather general assumptions about the transition probability $Q(\cdot, \cdot)$ and the function $g(\cdot)$) the convergence of the sequence of probability measures $P_j(\cdot)$ to a probability measure concentrated at the vicinity of the global maximizer of the function $g(\cdot)$.

The limiting distribution First, we introduce some notation that will be used below. Let $\mathscr{X}$ be a compact metric space (the main special case is where A is a bounded subset of $\mathbb{R}^d$); $\mathscr{B}$ be the σ-algebra of Borel subsets of $\mathscr{X}$; $\mathscr{M}$ be the space of finite signed measures, i.e. regular (countable) additive functions on $\mathscr{B}$ of bounded variation; $\mathscr{M}_+$ be the set of finite measures on $\mathscr{B}$ ($\mathscr{M}_+$ is a cone in the space $\mathscr{M}$); $\mathscr{M}^+$ be the set of probability measures on $\mathscr{B}$ ($\mathscr{M}^+ \subset \mathscr{M}_+$); $C_+(\mathscr{X})$ be the set of continuous non-negative functions on $\mathscr{X}$ ($C_+(\mathscr{X})$ is a cone in $C(\mathscr{X})$, the space of continuous functions on $\mathscr{X}$); $C^+(\mathscr{X})$ be the set of continuous positive functions on $\mathscr{X}$ ($C^+(\mathscr{X})$ is the interior of the cone $C_+(\mathscr{X})$); and a function K: $\mathscr{X} \times \mathscr{B} \to \mathbb{R}$ be such that $K(\cdot, U) \in C_+(\mathscr{X})$ for each $U \in \mathscr{B}$ and $K(x, \cdot) \in \mathscr{M}_+$ for each $x \in \mathscr{X}$. The analytical form of K may be unknown, but it is required that for any $x \in \mathscr{X}$ a method be known for evaluating realizations of a non-negative random variable $y(x)$ such that $\mathrm{E}y(x) = g(x) = K(x, \mathscr{X})$, $\mathrm{var}\, y(x) \leqslant \sigma^2 < \infty$, and of sampling the probability measure $Q(x, dz) = K(x, dz)/g(x)$ for all $x \in$

$\{x \in \mathscr{X} : g(x) \neq 0\}$. Denote by $\mathscr{K}$ the linear integral operator from $\mathscr{M}$ to $\mathscr{M}$ by $\mathscr{K}\nu(\cdot) = \int \nu(dx)K(x, \cdot)$. The conjugate operator $\mathscr{L} = \mathscr{K}^* : C(\mathscr{X}) \to C(\mathscr{X})$ is defined as follows:

$$\mathscr{L}h(\cdot) = \int h(x)K(\cdot, dx). \tag{3.3.3}$$

As it is known from the general theory of linear operators (see [11]), any bounded linear operator mapping from a Banach space into $C(\mathscr{X})$ can be represented by (3.3.3) and $\|\mathscr{L}\| = \|\mathscr{K}\| = \sup g(x)$. Moreover, the operators $\mathscr{K}$ and $\mathscr{L}$ are completely continuous in view of compactness of $\mathscr{X}$ and continuity of $K(\cdot, U)$ for all $U \in \mathscr{B}$.

The theory of linear operators in a space with cone (see, e.g. [25]) implies that a completely continuous and strictly positive operator $\mathscr{L}$ has eigen-value λ that is maximal in absolute value and positive, simple and at least one eigen-element (eigen-vector) belonging to the cone corresponds to it; the conjugate operator $\mathscr{L}^*$ has the same properties. In the present case, the operator $\mathscr{L}$ is determined by (3.3.3). It is strictly positive if for any non-zero function $h \in C_+(\mathscr{X})$, there exists $m = m(h)$ that $\mathscr{L}^m h(\cdot) \in C^+(\mathscr{X})$ where $\mathscr{L}^m$ is the operator with kernel $\int \ldots \int K(\cdot, dx_1)K(x_1, dx_2) \ldots K(x_{m-1}, \cdot)$. Thus, if the operator $\mathscr{L} = \mathscr{K}^*$ is strictly positive (which is assumed to be the case), the maximal in absolute value eigen-value λ of $\mathscr{K}$ is simple and positive; a unique eigen-measure $R(\cdot)$ in $\mathscr{M}^+$ defined by $\lambda R(dx) = \int R(dz)g(z)Q(z, dx)$ corresponds to this λ, and λ itself is expressed in terms of the measure $R(\cdot)$ as $\lambda = \int g(x)R(dx)$. This yields that if $\lambda \neq 0$, then $R(\cdot)$ is a unique in $\mathscr{M}^+$ eigen-measure of $\mathscr{K}$ if and only if $R(\cdot)$ is a unique in $\mathscr{M}^+$ solution of the integral equation

$$R(dx) = \left[\int g(z)R(dz)\right]^{-1} \int R(dz)g(z)Q(z, dx); \tag{3.3.4}$$

We shall call this solution 'probabilistic solution'.

The following theorem proved in [51] is the key statement relating Algorithm 1 to the probability measure $R(\cdot)$, the solution of the integral equation (3.3.4).

Theorem 3.3 *Let $Q(z, dx) \geqslant c\,dx$ for almost all $z \in \mathscr{X}$ where c is some positive constant. Then, under some assumption of regularity on f and $\mathscr{X}$: (1) for any $n = 1, 2, \ldots$ the random vectors $a_j = (x_1^{(j)}, \ldots, x_n^{(j)})$, $j = 1, 2, \ldots$, constitute a homogeneous Markov chain with stationary distribution $W_n(dx_1, \ldots, dx_n)$, the random vectors with this distribution are symmetrically dependent; (2) for any $\varepsilon > 0$ there exists $n_* \geqslant 1$ such that for $n \geqslant n_*$ the marginal distributions $W_{(n)}(\cdot) = W_n(\cdot, \mathscr{X}, \ldots, \mathscr{X})$ differ in variation from $R(\cdot)$ at most by ε.*

For the proof, see Theorem 5.3.1 in [51].

Another way of relating Algorithm 1 to $R(\cdot)$, the solution to (3.3.4), would be to consider $R(\cdot)$ as the limit of the marginal probability measures $R_j(\cdot)$. More precisely (see [51], Corollary 5.2.2), in a general case where $g(\cdot)$ is evaluated

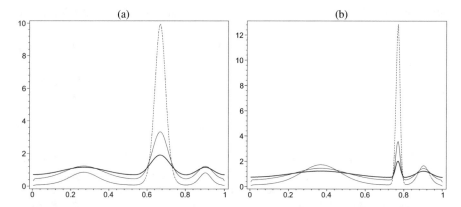

Fig. 3.9 Iterations of the densities of the probability measures $R_j(\cdot)$ defined by (3.3.5) with $g(x) = \exp\{-f(x)\}$, $x \in A = [0, 1]$. The function $f(\cdot) = f_k(\cdot)$ is defined with $k = 3$ and $k = 12$

with random errors, the unconditional distributions of random points $x_i^{(j+1)}$ of Algorithm 1 converge in variation (as $n \to \infty$) to the limiting distributions $R_{j+1}(\cdot)$ satisfying

$$R_{j+1}(dx) = \left[\int R_j(dz)g(z)\right]^{-1} \int R_j(dz)g(z)Q(z, dx) \qquad (3.3.5)$$

(where $R_1(\cdot) = P_1(\cdot)$). These distributions, in turn, converge (as $j \to \infty$) to $R(\cdot)$, the probabilistic solution of the equation (3.3.4).

Convergence of the probability measures $R_j(\cdot)$: an example In Fig. 3.9 we illustrate the iterations of the densities $p_j(\cdot)$ corresponding to the probability measures $R_j(\cdot)$ defined by (3.3.5). Here $g(x) = \exp\{-f(x)\}$ for $x \in [0, 1]$ and $g(x) = 0$ otherwise; the transition density $q = q_\beta$ has the form (3.3.6) with $\beta = 1$ and $\phi(u) = \frac{1}{\sqrt{2\pi}} \exp(-u^2/2)$; the initial density $p_1(\cdot)$ is $g(x)/\int g(u)du$, $x \in [0, 1]$.

The following densities (restricted to the interval $[0, 1]$) are plotted: $p_1(\cdot)$ (thick solid line), $p_2(\cdot)$ (thin solid line) and $p_5(\cdot)$ (dotted line). Figure 3.9 illustrates that in the cases considered, the measures $R_j(\cdot)$ quickly converge to the vicinity of the global minimizer x_*. The limiting measure in both cases is concentrated at the vicinity of the global minimizer x_*; see below.

Studying the closeness between the probabilistic solution of (3.3.4) and the δ-measure concentrated at the global maximizer of $g(\cdot)$ We have shown above that the marginal distributions of the random points generated by Algorithm 1 (with $f_j(\cdot) = g(\cdot)$ and $Q_j(\cdot, \cdot) = Q(\cdot, \cdot)$) converge to $R(\cdot)$, the probabilistic solution to the equation (3.3.4). The next problem is to study closeness between

$R(\cdot)$ (the probabilistic solution of the equation (3.3.4)) and $\varepsilon^*(dx)$, the δ-measure concentrated at the global maximizer of the function $g(\cdot)$.

We assume that the transition probabilities $Q(x, dz) = Q_\beta(x, dz)$ weakly converge to $\varepsilon_x(dz)$ as $\beta \to 0$ (here $\varepsilon_x(dz)$ is the δ-measure concentrated at x).

In order to relieve the presentation of unnecessary details, assume that $\mathscr{X} = \mathbb{R}^d$ and that the transition densities $q_\beta(x, z)$ are chosen with $\beta_j = \beta$, i.e.

$$q_\beta(x, z) = \beta^{-d} C \varphi((z - x)/\beta). \tag{3.3.6}$$

The probability density $\beta^{-d} \varphi((z - x)/\beta)$ can be considered as a smoothing kernel in $\mathbb{R}^d$. The following proposition states that if β is sufficiently small ($\beta > 0$), then under natural assumptions concerning the objective function, the majority of points generated by Algorithm 1 will be concentrated at the vicinity of the global minimizer of $f(\cdot)$.

Proposition 3.1 (see Lemma 5.3.1 in [51]) *Let the transition density $q = q_\beta$ have the form (3.3.6), where φ is a continuously differentiable density on $\mathbb{R}^d$, $\int \|x\| \varphi(x) dx < \infty$, $g(\cdot)$ be positive, satisfy the Lipschitz condition, attain the global maximum at a unique maximizer x_*, and $g(x) \to 0$ for $\|x\| \to \infty$. Then for any $\varepsilon > 0$ and $\delta > 0$, there exists $\beta > 0$ such that $R(B(x_*, \delta)) \geqslant 1 - \varepsilon$, where $R(\cdot)$ is the probabilistic solution of (3.3.4).*

Heuristically, Proposition 3.1 can be illustrated using the following reasoning. In the case studied, $R(dx)$ has a density $p(x)$ that may be obtained as the limit (for $j \to \infty$) of recurrent approximations $p_{j+1}(x) = s_{j+1} \int p_j(z) g(z) \varphi_\beta(x - z)(dz)$, where $\varphi_\beta(x) = \beta^{-d} \varphi(x/\beta)$, $s_{j+1} = 1 / \int \int p_j(z) g(z) dz$. This updating means that $p_{j+1}(\cdot)$ are kernel estimators of the densities $s_{j+1} p_j(\cdot) g(\cdot)$, where the parameter β is often called 'window width'. One can anticipate that for small β the asymptotic behaviour of densities $p_j(\cdot)$ should not differ much from that of densities proportional to $g^j(x)$; at the same time, as $j \to \infty$, the sequence of the distributions with these densities converges to $\varepsilon^*(dx)$.

Numerical calculations confirm the fact that in the problems resembling realistic ones (for not 'too bad' functions f), the measures $R = R_\beta$ explicitly tend, for small β, to concentrate mostly within a small vicinity of the global maximizer x_* (or the maximizers). Moreover, this tendency typically manifests itself already for not very small β (e.g. $\beta \simeq 0.2$, under the assumption that the covariance matrix of the distribution with the density φ is the identity matrix).

Extension to genetic-type algorithms In the genetic-type algorithms, each descendant has two parents. Assume that $f_j(\cdot, \cdot) = g(\cdot, \cdot)$ for some function $g(\cdot, \cdot)$ and that the transition probabilities $Q_j(y, z, \cdot)$ are such that $Q_j(y, z, \cdot) = Q(y, z, \cdot)$ for all y, z (i.e. the updating rule is the same at each iteration). Then the the the main formulae of this section can be generalized to this case. For example, the updating formula (3.3.5) becomes

$$R_{j+1}(dx) = \left[\int g(y,z) R_j(dy) R_j(dz) \right]^{-1} \int g(y,z) R_j(dy) R_j(dz) Q(y,z,dx), \quad (3.3.7)$$

and the integral equation (3.3.4) for the limiting measure $R(\cdot) = \lim_{j \to \infty} R_j(\cdot)$ will have the form

$$R(dx) = \left[\int g(y,z) R(dy) R(dz) \right]^{-1} \int g(y,z) R(dy) R(dz) Q(y,z,dx). \quad (3.3.8)$$

Note that unlike (3.3.4), which is a linear integral equation, the equation (3.3.8) is nonlinear. This nonlinearity, however, is only quadratic, which is one of the easiest nonlinearities to deal with. We leave to interested researchers in the problem of studying the convergence of the sequence of distributions (3.3.7) to the solution of (3.3.8) and the problem of investigating properties (including existence and uniqueness) of the probabilistic solution/s of the equation (3.3.8).

On construction of basic genetic algorithms in continuous problems By following the standard literature [20, 31, 36], let us consider construction of a typical genetic algorithm in an optimization problem with continuous (interval-scale) variables. Let $x \in [0, 1]$ be one of the input variables. It is coded as n binary digits a_i ($i = 1, \ldots, n$) by taking the binary representation of $x = \sum_{i=1}^{\infty} a_i 2^{-i}$ and keeping the first n binary digits only. After performing such operation with all continuous variables, the original continuous problem is discretised into 0, 1-variables and solved with the usual techniques of mutation (which can be well-defined) and recombination, or crossover. While a crossover for binary variables makes perfect sense, by looking at its effect in the original (interval-scaled) variables, the operation becomes senseless as it cannot be matched with the Euclidean distance in the original space; that is, the result of crossover of two points does not make clear sense from the viewpoint of Euclidean geometry. Note, however, that the community of genetic algorithms has evolved and started to realize difficulties like this. Some of the recent books on genetic and, more generally, evolutionary algorithms show a much better understanding of how different heuristics agree with the underlying geometry of the space; see, e.g. [2, 4, 40].

3.4 Bibliographic Notes

The main methodological and especially theoretical advances in GRS have been made by the first co-author more than 30 years ago and published in [49, 51]. Since that time, there was no clear progress in the theory of GRS (here we are taking aside the theory of Markov global random search [42–44]), and the main progress in methodology of GRS is associated with meta-heuristical arguments; see Sect. 3.1.4. Note however that there are different (but not contradictory) viewpoints towards the methodology of GRS; see, e.g. [2, 22, 29, 34, 45].

Section 3.1 mostly follows [49, 51, 53]. Results of Sect. 3.1.2 use [54] and [56]. Sect. 3.1.3 follows [49, 51, 53] but similar results can be found in [22, 33, 41]. For many different meta-heuristics mentioned in Sect. 3.1.4, see [1, 3, 10, 12, 19, 37]. The view expressed in Sect. 3.1.4 is entirely personal.

Sections 3.2 and 3.3 contain a carefully selected material from [49, 51, 53]. In particular, these references contain much more material on the important topic of statistical inferences about f_*, briefly discussed in Sect. 3.2. Section 3.2.4 uses results of [32]. Among the ideas discussed in Sect. 3.3, the concept illustrated in Fig. 3.9 seems to be most important.

Applications of GRS are countless; they are considered in many publications (see, e.g. [2, 4, 22, 34]) as well as almost any literature on the meta-heuristics. I would also like to attract attention of specialists in global optimization to two important statistical applications: construction of optimal designs [8, 9, 35] and structured low-rank matrix approximation [13–16]; these two areas are not sufficiently noticed by the specialists in global optimization. In both of them, the global optimization problems arising are very difficult, but, as shown in [9, 13–15, 35], clever reformulations allow to create intelligent and very efficient optimization strategies.

References

1. A. Auger, B. Doerr (eds.), *Theory of Randomized Search Heuristics: Foundations and Recent Developments* (World Scientific, Singapore, 2011)
2. R. Battiti, M. Brunato, The lion way: machine learning plus intelligent optimization. *LIONlab*, University of Trento, Italy, 94 (2014)
3. Z. Beheshti, S.M.H. Shamsuddin, A review of population-based meta-heuristic algorithms. Int. J. Adv. Soft Comput. Appl. **5**(1), 1–35 (2013)
4. J. Branke, *Evolutionary Optimization in Dynamic Environments* (Springer, Berlin, 2012)
5. P. Cooke, Optimal linear estimation of bounds of random variables. Biometrika **67**, 257–258 (1980)
6. L. De Haan, L. Peng, Comparison of tail index estimators. Statistica Neerlandica **52**(1), 60–70 (1998)
7. K. Deb, An efficient constraint handling method for genetic algorithms. Comput. Methods Appl. Mech. Eng. **186**(2–4), 311–338 (2000)
8. H. Dette, A. Pepelyshev, A. Zhigljavsky, Optimal design for linear models with correlated observations. Ann. Stat. **41**(1), 143–176 (2013)
9. H. Dette, A. Pepelyshev, A. Zhigljavsky, Optimal designs in regression with correlated errors. Ann. Stat. **44**(1), 113 (2016)
10. K.-L. Du, M. Swamy, *Search and Optimization by Metaheuristics* (Birkhauser, Basel, 2016)
11. N. Dunford, J.T. Schwartz, *Linear Operators. Part I* (Wiley, New York, 1988)
12. M. Gendreau, J.-Y. Potvin, others (eds.), *Handbook of Metaheuristics*, Vol. 2 (Springer, New York, 2010)
13. J. Gillard, K. Usevich, Structured low-rank matrix completion for forecasting in time series analysis. Int. J. Forecast. **34**(4), 582–597 (2018)
14. J. Gillard, A. Zhigljavsky, Optimization challenges in the structured low rank approximation problem. J. Global Optim. **57**(3), 733–751 (2013)

15. J. Gillard, A. Zhigljavsky, Stochastic algorithms for solving structured low-rank matrix approximation problems. Commun. Nonlinear Sci. Numer. Simul. **21**(1–3), 70–88 (2015)
16. J. Gillard, A. Zhigljavsky, Weighted norms in subspace-based methods for time series analysis. Numer. Linear Algebra Appl. **23**(5), 947–967 (2016)
17. J. Gillard, A. Zhigljavsky, Optimal directional statistic for general regression. Stat. Probab. Lett. **143**, 74–80 (2018)
18. J. Gillard, A. Zhigljavsky, Optimal estimation of direction in regression models with large number of parameters. Appl. Math. Comput. **318**, 281–289 (2018)
19. F.W. Glover, G.A. Kochenberger (eds.), *Handbook of Metaheuristics* (Springer, New York, 2006)
20. D.E. Goldberg, *Genetic Algorithm in Search, Optimization and Machine Learning* (Addison-Wesley, Reading, 1989)
21. E.M.T. Hendrix, O. Klepper, On uniform covering, adaptive random search and raspberries. J. Global Optim. **18**(2), 143–163 (2000)
22. E.M.T. Hendrix, B.G.-Tóth, *Introduction to Nonlinear and Global Optimization*, Vol. 37 (Springer, New York, 2010)
23. J. Hooker, Testing heuristics: we have it all wrong. J. Heuristics **1**, 33–42 (1995)
24. N. Ketkar, Stochastic gradient descent. In: *Deep Learning with Python*, pp 113–132 (Springer, New York, 2017)
25. M.A. Krasnosel'skij, J.A. Lifshits, A.V. Sobolev, *Positive Linear Systems* (Heldermann Verlag, Berlin, 1989)
26. P. Kulczycki, S. Lukasik, An algorithm for reducing the dimension and size of a sample for data exploration procedures. Int. J. Appl. Math. Comput. Sci. **24**(1), 133–149 (2014)
27. V.B. Nevzorov, *Records: Mathematical Theory* (American Mathematical Society, Providence, 2001)
28. J. Noonan, A. Zhigljavsky, Appriximation of the covering radius in high dimensions (2021, in preparation)
29. P. Pardalos, A. Zhigljavsky, J. Žilinskas, *Advances in Stochastic and Deterministic Global Optimization* (Springer, Switzerland, 2016)
30. N.R. Patel, R.L. Smith, Z.B. Zabinsky, Pure adaptive search in Monte Carlo optimization. Math. Program. **43**(1–3), 317–328 (1989)
31. M. Pelikan, *Hierarchical Bayesian Optimization Algorithm* (Springer, Berlin, Heidelberg, 2005)
32. A. Pepelyshev, A. Zhigljavsky, A. Žilinskas, Performance of global random search algorithms for large dimensions. J. Global Optim. **71**(1), 57–71 (2018)
33. J. Pintér, Convergence properties of stochastic optimization procedures. Optimization **15**(3), 405–427 (1984)
34. J. Pinter, *Global Optimization in Action* (Kluwer Academic Publisher, Dordrecht, 1996)
35. L. Pronzato, A. Zhigljavsky, Algorithmic construction of optimal designs on compact sets for concave and differentiable criteria. J. Stat. Plann. Inference **154**, 141–155 (2014)
36. C.R. Reeves, J.E. Rowe, *Genetic Algorithms: Principles and Perspectives* (Kluwer, Boston, 2003)
37. C. Ribeiro, P. Hansen (eds.), *Essays and Surveys in Metaheuristics* (Springer, New York, 2012)
38. A.H.G. Rinnooy Kan, G.T. Timmer, Stochastic global optimization methods. Part I: clustering methods. Math. Program. **39**(1), 27–56 (1987)
39. S.K. Sahu, A. Zhigljavsky, Self-regenerative Markov chain Monte Carlo with adaptation. Bernoulli **9**(3), 395–422 (2003)
40. D. Simon, *Evolutionary Optimization Algorithms* (Wiley, Chichester, 2013)
41. F. Solis, R. Wets, Minimization by random search techniques. Math. Oper. Res. **6**(1), 19–30 (1981)
42. D. Tarłowski, On the convergence rate issues of general Markov search for global minimum. J. Global Optim. **69**(4), 869–888 (2017)
43. A.S. Tikhomirov, On the convergence rate of the simulated annealing algorithm. Comput. Math. Math. Phys. **50**(1), 19–31 (2010)

44. A. Tikhomirov, T. Stojunina, V. Nekrutkin, Monotonous random search on a torus: integral upper bounds for the complexity. J. Stat. Plann. Inference **137**(12), 4031–4047 (2007)
45. A. Törn, A. Žilinskas, *Global Optimization* (Springer, Berlin, 1989)
46. W. Tu, W. Mayne, Studies of multi-start clustering for global optimization. Int. J. Numer. Meth. Eng. **53**, 2239—2252 (2002)
47. P. Van Laarhoven, E. Aarts, *Simulated Annealing: Theory and Applications* (Kluwer, Dordrecht, 1987)
48. Z.B. Zabinsky, *Stochastic Adaptive Search for Global Optimization* (Kluwer, Boston, 2003)
49. A. Zhigljavsky, *Mathematical Theory of Global Random Search* (Leningrad University Press, Leningrad, 1985). in Russian
50. A. Zhigljavsky, Branch and probability bound methods for global optimization. Informatica **1**(1), 125–140 (1990)
51. A. Zhigljavsky, *Theory of Global Random Search* (Kluwer, Dordrecht, 1991)
52. A. Zhigljavsky, E. Hamilton, Stopping rules in k-adaptive global random search algorithms. J. Global Optim. **48**(1), 87–97 (2010)
53. A. Zhigljavsky, A. Žilinskas, *Stochastic Global Optimization* (Springer, New York, 2008)
54. R. Zieliński, A statistical estimate of the structure of multi-extremal problems. Math. Program. **21**, 348–356 (1981)
55. A. Žilinskas, A. Zhigljavsky, Branch and probability bound methods in multi-objective optimization. Optim. Lett. **10**(2), 1–13 (2016)
56. A. Žilinskas, J. Gillard, M. Scammell, A. Zhigljavsky, Multistart with early termination of descents. J. Global Optim. 1–16 (2019). https://doi.org/10.1007/s10898-019-00814-w

Printed in the United States
By Bookmasters